JEFF ROTMAN

UNDERSEA JOURNEY

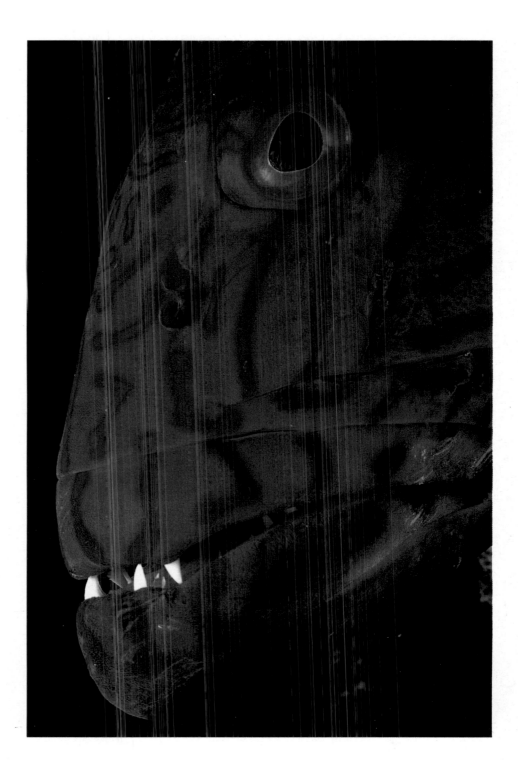

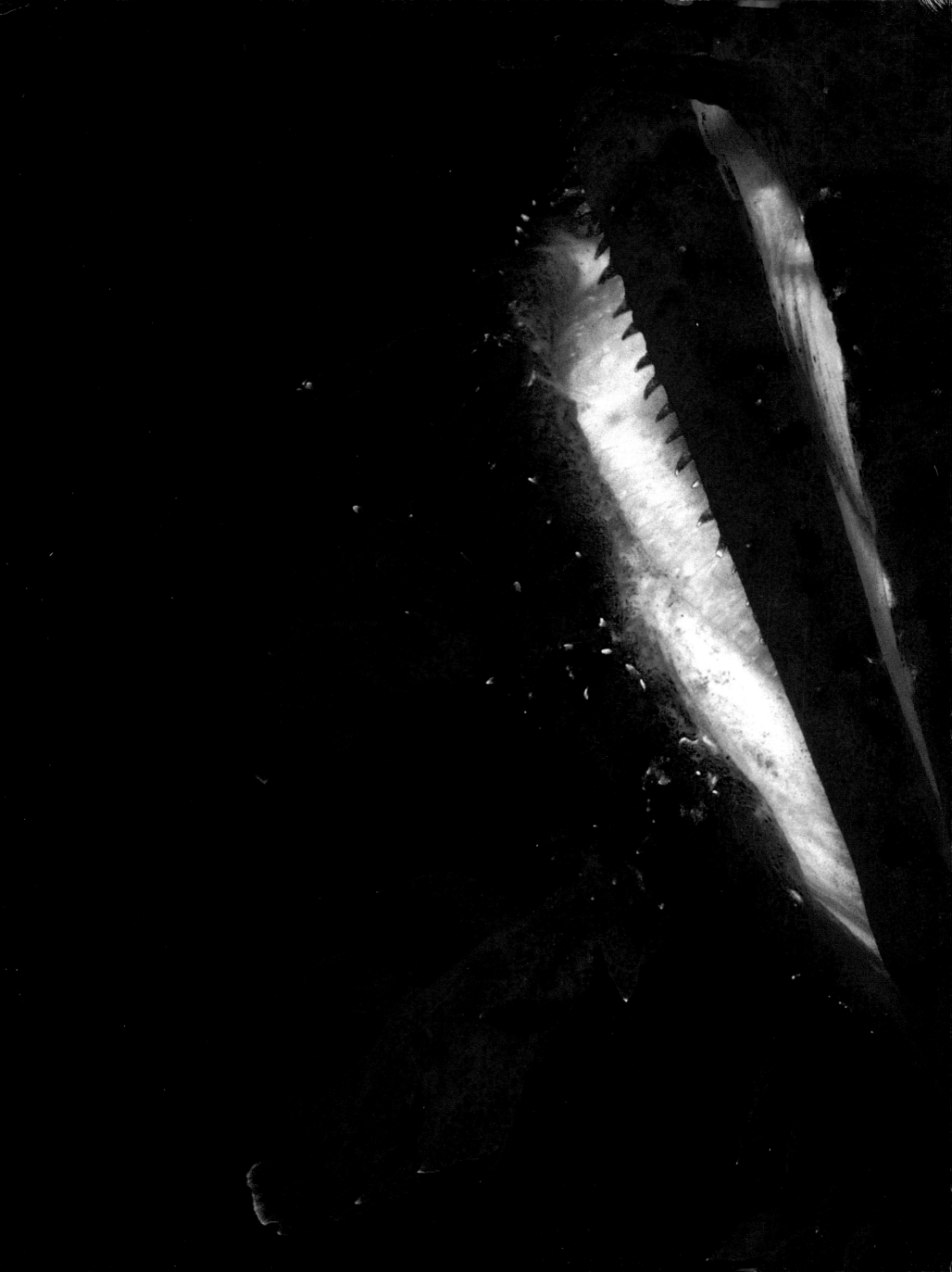

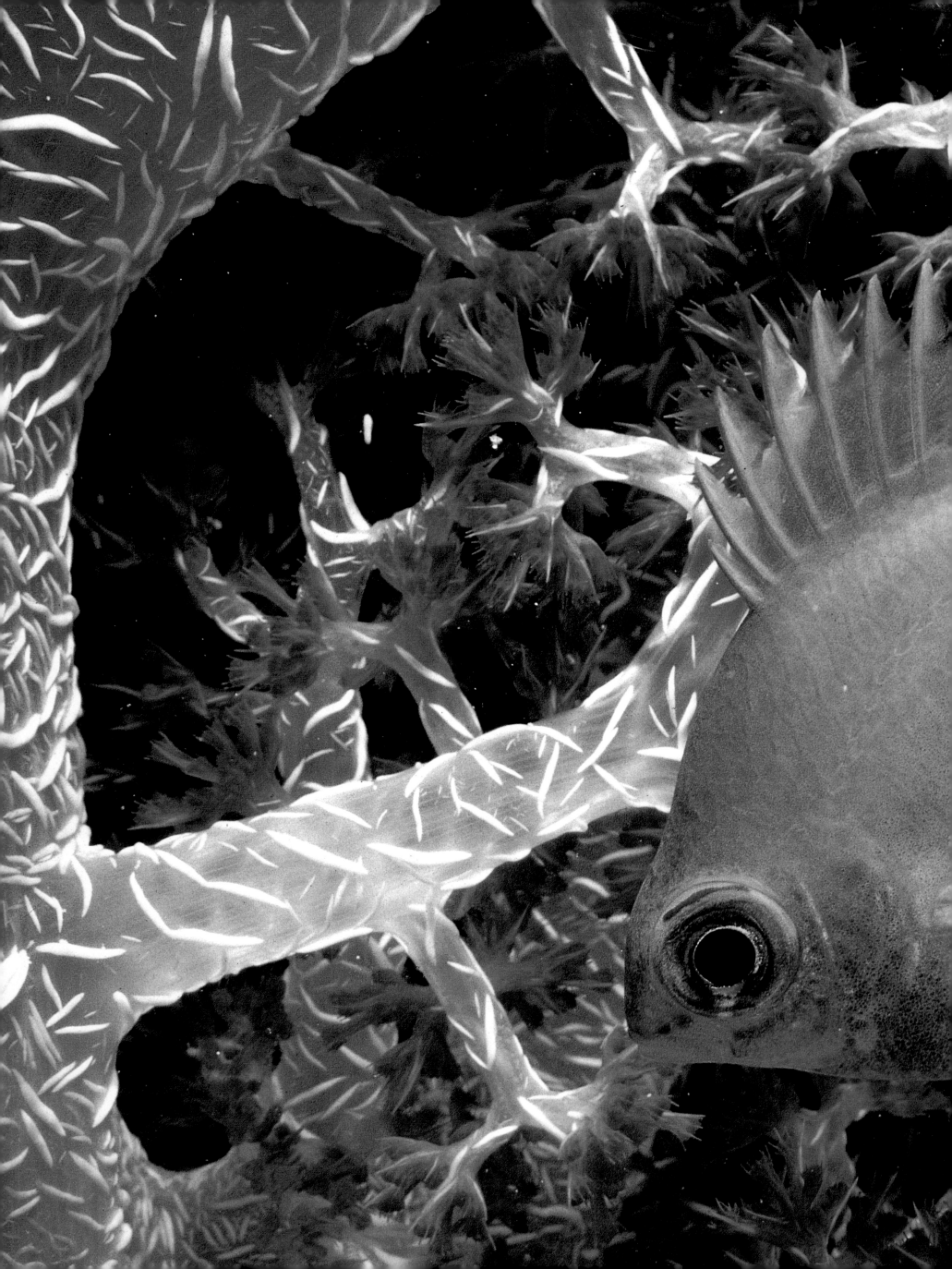

UNDERSEA JOURNEY

Text and Photographs
Jeffrey L. Rotman

Editor
Valeria Manferto

Editorial coordination
Laura Accomazzo

Designer
Patrizia Balocco
Anna Galliani

Biological consultant
Angelo Mojetta

CONTENTS

This edition published in 1995
by Smithmark Publishers, a
division of U.S. Media
Holdings, Inc.,
16 East 32nd Street,
New York, NY 10016.

SMITHMARK books are
available for bulk purchase, for
sales promotion and premium
use. For details write or call the
manager of special sales,
SMITHMARK Publishers, 16
East 32nd Street, New York, NY
10016; (212) 532-6600.

Produced by: White Star S.r.l.
Via Candido Sassone, 22/24 -
13100 Vercelli, Italy.

ISBN: 0-8317-1038-1

Printed in the month of July
1995 by Canale, Turin (Italy).
Color separation Fotolito
Garbero, Turin.

10 9 8 7 6 5 4 3 2 1

The Author would like to thank:
Jane, Bob, Jen and John Altman, Peter Arnold,
Ken Beck, Cecile and Max Benjamin, Mary
Cerullo, Howard Chapnick, Natasha Chassagne,
Max Chavanne, Michelle and Stuart Cove, Bob
Cranston, Bill Curtsinger, Isabella Delafosse,
Jerome Delafosse, Fred Dion and Underwater
Photo Tech, Embarak of Sinai, Rodney Fox,
David Friedman, Asher Gal, Amos Goren,
Mohammad Hagrass, Willy Halpert, Yair Harrel,
Paul Humann, Venita Kaleps, Margaret Kaplan,
Maria Lane, Joe Levine, Oren Lifshitz, Barbara
London, Brian Lyons, Claudine Maugendre,
Angelica and Ernst Meier, Urs Möckli, Koji
Nakamura, Lello Piazza, Sylvie Rebbot, Repro
Images, Ben Rose, Howard Rosenstein, Dana,
Adam, and Galit Rotman, Rene and Nate
Rotman, Kathy Ryan, Mel Scott, Maria Steurer,
Sal Tecci, Francis Toribiong, Stan Waterman,
Neal Watson.

1 The sharp teeth of this grouper *(Plectropomus oligacanthus)* reveal its predatory nature.
Kimbe Bay,
Papua New Guinea.
Depth: 7 meters.

2-3 This bearded visage of a sea raven *(Hemitripterus americanus)*, taken 18 years ago, never seems to lose its power.
Gloucester, Massachusetts.
Depth: 7 meters.

4-5 A golden damselfish *(Amblyglyphidodon aureus)* resting in the branches of an alcyonarian at 3:00 a.m.
Walindi Bay, Papua New Guinea. Depth 9 meters.

6-7 Alcyonarians *(Dendronephthya sp.)* are among the most spectacularly colored of all marine invertebrates.
Kimbe Bay, Papua New Guinea.
Depth 13 meters.

8 An ethereal white
nudibranch (*Dirona
Alboljneata*) fluctuates in
the darkness of the night.
Puget Sound,
Pacific Ocean.
Depth: 10 meters.

8-9 When people tell me
that the Mediterranean Sea
is hardly worth diving I
pull out this photograph
and tell them that every sea
has its jewel: a nudibranch
(*Flabellina affinis*)
undulates displaying its
poison-tipped cerata.
Ustica Island,
Mediterranean Sea.
Depth: 8 meters.

10-11 Caribbean reef
sharks (*Carcharhinus
perezi*) can be approached
safely if attempted with
patience-breath slowly
and remain absolutely
still - curiosity usually
wins out over
apprehension.
Walkers Cay, Bahamas.
Depth: 15 meters.

12-13 Inside a school of
barracuda (*Sphyraena sp.*).
Kimbe Bay, Papua New
Guinea.
Depth: 15 meters.

INTRODUCTION

For the past twenty years, I have chronicled the sea. I have traveled the world's oceans in an attempt to discover its many well-kept secrets and record them on film. Over thousands of dives, some too deep, others that pushed the limits of hypothermia, all rewarding, I have worked toward a single purpose — to photograph the secrets of the sea during a continuous undersea journey. I first took the plunge in 1973, after a friend persuaded me that the chilly North Atlantic surrounding my New England home just might prove interesting. It also proved frigid and unforgiving. The body of water that borders most of New England and Maritime Canada is known as the Gulf of Maine, colder and richer than the rest of the Atlantic Ocean because of the many rivers that feed nutrients into it. Despite having worked as a lifeguard for many years, my swimming skills needed upgrading in order for me to become comfortable in this hostile environment. I realized that I could not become dependent on scuba gear if I was going to deal calmly with emergencies in these murky, disorienting depths. I had to teach myself to abandon both the surface and my scuba gear and become a breath-holding free-diver. With practice, I disciplined myself to hold my breath and descend to 9 meters and spend a minute there. During that minute, the ocean in front of my face mask belonged only to me, without the distractions of bubbles or breathing gear. During my first few free-diving attempts, I saw little but kelp and other seaweeds. My thoughts and energy were focused on my diving and, most importantly, on returning to the surface. Gradually, though, free-diving gave me the confidence I needed to expand my limits as a scuba diver and to dispel the psychological monsters that stalked me beneath the sea. Giant octopuses and sharks did not lurk behind every boulder. In fact, it would be years before I would meet either one under the sea. The diving began to feel more natural. I was no longer preoccupied with the mechanics of diving. It became second nature. When this happened, only then did I begin to see and feel the ocean. In the early 1970's, diving in New England usually consisted of hunting that gourmet delight, the Maine lobster, which was still plentiful there, or chasing schools of fish with a speargun into the shipwrecks that abounded in these treacherous waters. I chose instead to dive armed with a camera. At that time I was a science teacher in Cambridge, Massachusetts. My students were underprivileged city kids who had never heard of a sea urchin or handled a seastar. Some had never been to the ocean, even though their city is nearly surrounded by the sea. As their reward for not disrupting class, I would haul out my

14-15 Held in the embrace of its poisonous protector, the anemone, this clownfish *(Amphiprion sp.)* can relax. Kesebekuu Pass Reef, Palau Islands. Depth: 8 meters.

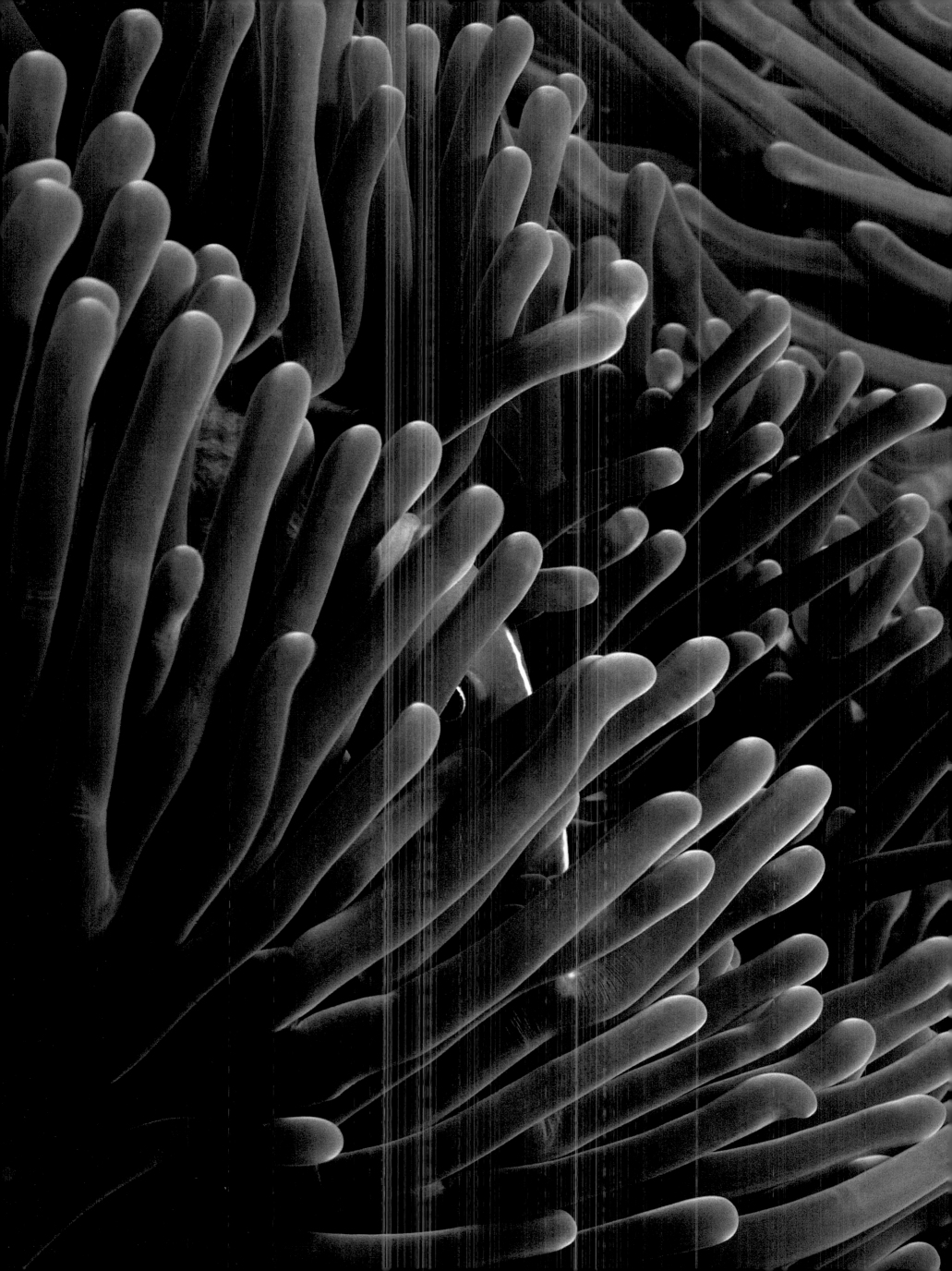

slides and introduce my students to their neighbors off the coast. They hung on every word and exclaimed at every photo. Their enthusiasm was contagious. I abandoned the textbook.

The sea, especially the North Atlantic, became the focus of my teaching. I took my students to Gloucester, along the rocky coast. I would dive for all manner of strange and bizarre marine life while they waited on shore.

When I surfaced with a bag full of "goodies" I felt like Santa Claus as my students crowded around excitedly for a closer look.

In 1976, I met Jacques Cousteau. I asked him to name his favorite dive place in the world. On his advice, I packed my bags and headed for the Middle East and the Red Sea. I would not be disappointed. That first dive off the sandy coast of Eilat, Israel, was mind-boggling. My dive companion and I were alone in an alien world of indescribable beauty. It was as if an untouched fairy land had been created for our appreciative eyes alone. I remember descending 10 meters and finding a coral outcrop the size of a microwave oven. For the next hour, I abandoned my cameras on the sandy bottom and surveyed this small patch of reef in amazement. A cloud of gold and red anthias hovered just above a coral boulder. I spied a coral grouper that had settled next to a pair of moray eels.

I discovered three different types of shrimp within that small space. One variety had set up housekeeping within the poisonous tentacles of a sea anemone in harmony with a pair of resident clownfish. Tucked deep inside the coral crevices were sea urchins, feather star crinoids, and basket stars, which would only make their appearance later, under the cover of darkness. A pajama nudibranch slowly inched its way along the underside of a bright red finger sponge. It was scraping off a microscopic meal with its radula, a file-like tongue. Two *Tridacna* clams had opened their shells wide enough to allow a free exchange of water from which they would filter their own planktonic meals. Toward the end of that first hour I finally dared to touch the coral. Some were hard as rock; others were barely more substantial than the water that filled them. As my hand moved over the coral, I felt something which was neither hard nor delicate; it had a supple elasticity that recoiled at my touch. I was barely able to make out the outline of an octopus that had stood its ground as I approached, evidently confident that its camouflage would allow it to go undiscovered. I experienced a feeling of breathlessness, and realized it was not due to my surroundings, but to my dwindling air supply. Still in a trance, I ascended through a school of sergeant major fish that seemed to think my hair was edible.

16-17 Beneath a crinoid you can occasionally find a clingfish *(Discotrema crinophila)* that lives symbiotically among its arms. The matching coloration of the host crinoid makes the fish very difficult to spot. Kimbe Bay, Papua New Guinea. Depth: 9 meters.

16 bottom and 17 bottom Crinoids are striking and unusual relatives of more familiar starfish and sea urchins. The most common types of crinoids, which are often (though not always) nocturnal, carry long, feathery appendages that sweep through the water to gather plankton. Walindi Bay, Papua New Guinea. Depth: 8 meters.

18 top The moot colored sculpin is hard and horny skinned. It lives in a rough and rugged environment - the first 10 meters of the blasting Atlantic shore-an environment that is harsh beyond belief. Bass Rocks, Gloucester, Massachusetts. Depth: 7 meters.

On the surface, I unleashed the war hoops of excitement I had had to contain underwater. Wow, how could this be! It was incredible, unbelievable... I had become a fish! I returned to the same area, dive after dive, sometimes with my camera, sometimes without. I couldn't resist the allure of the Red Sea reefs.

When I returned to Boston and developed my film, I was delighted with the results. My photos had captured the colors and complexity of the reef. My awe had been transferred to film. How much easier it seemed to get good photographs in a warm, clear, colorful sea than in the cold, dark waters of the Atlantic where I had apprenticed my underwater shooting. But New England diving still thrilled me. The challenge of its difficult conditions definitely made me be a better diver. In New England we have a name for it; we call it gorilla diving.

Soon after, I sent my photos to a small diving magazine, along with a text about the bizarre marine life of New England waters. It was accepted and published. My name and photos in print! I was hooked! I packaged my material on the Red Sea, sent it off, and it met with the same success. Seeing my photographs published was the inspiration I needed to throw myself into diving the Atlantic year-round.

Diving in the North Atlantic Ocean is always cold, but in the dead of winter, the water temperature may be 33 F (+ 1°C) and the air temperature well below freezing. When ice freezes on your eyebrows as you climb out of the water, you have to develop an accepting attitude towards being cold. You try not to focus on how cold you are in a relatively thin wet suit, but on how long you can maintain feeling in your fingers. But you can ward off the acknowledgement of the cold for only so long. Then it hits you as suddenly and as forcefully as slamming into a brick wall (or a wall of ice). The dive ends when the shivering becomes so severe you can't hold the camera still any longer. But there are many advantages to diving through the cold months. With a decrease in the hours of sunlight each day comes a dramatic drop in the plankton that impairs your visibility. Instead of swimming through pea soup, the water comes closer to being a clear broth. Perhaps the cleanest water I have ever seen was in New England in January, at a depth of 30 meters, beneath the second thermocline, the zone marking a sudden change in water temperature. The ocean was as clear as gin.

I was always much more aware of seasonal changes in New England waters. This ocean, like the land, had a clearly demarcated Spring, Summer, Fall, and Winter. With Spring came blooming plants, birth, and rapid growth.

In late Fall, migration, hibernation beneath the

18-19 A sea raven (*Hemitripterus americanus*) was well camouflaged in a bed of sea anemones until my flash brought color and graphics to this scene. The raven lies directly in front of a forward torpedo tube on the sunken submarine U-853. She was depth-charged 8 hours before the second world war ended - all hands lost. May she rest in peace. Block Island, Rhode Island.
Depth: 35 meters.

20 You are what you eat: this body builder, a pajama nudibranch *(Chromodoris quadricolor),* gets its striking coloration from its diet. Jackson Reef, Red Sea. Depth: 6 meters.

21 This emperor angelfish *(Pomacanthus imperator)* is easy to spot in a crowd. Marsa Baraka, Red Sea. Depth: 12 meters.

sand, and death marked the end of another yearly cycle beneath the waves. I observed the changes throughout each year. I was a student and the ocean was my respected teacher. In 1979, I was ready to tackle a new goal. I began work on my first book, *Beneath Cold Seas.* The stated purpose of this book was to "explore the cold temperate waters of North America."

I loaded my photo and dive gear into my car and, like millions before me, I headed west to California. Only I didn't stop at the shoreline. I found the *Truth,* owned and captained by the ever-resourceful Roy Hauser, to take me to the offshore kelp forests of California, one of the great underrated dive locations. *Truth* became my home that summer, as it steamed from one Channel Island to another, allowing me to see and film the many faces of California's rich kelp forests.

From there I migrated up the Pacific coast to Puget Sound and on to British Columbia. These cold waters offered a totally different marine flora and fauna from the kelp forests of southern California. Off Hornby Island in British Columbia I met my first giant Pacific octopus — what a creature! Stretched end to end, it would have filled a living room, but huddled inside its rocky den, it seemed no bigger than a breadbox. Sixteen years later I would return to the chilly waters of the Strait of Georgia to

work with this shy and reclusive giant again. For a week I followed a resident population of giant octopuses. I watched them stalk crabs and mollusks, scoop them into their weblike arms, tiptoe back to their dens, and toss the empty shells outside the door. I peeked at mother octopuses inside nearly-sealed caves as they nurtured their offspring. I knew that by the time they hatched, their mother would die of starvation. During these early years I was drawn to close-up macro photography. Learning to dive as I had in low visibility made close-up photography a necessity. In water so murky that images disappear within 2-3 meters, I had little choice. I focused my energy instead on deciphering the details of the marine life, their interactions with each other, their relationships of symbiosis, courtships, and predator and prey. As my macro photography progressed, I moved in even closer, concentrating on the colors and textures of my subjects: the ruby-red mouth of a sea anemone, the fleshy appendages hanging from the face of a yellow sea raven, the impossible shade of orange covering a sea peach.

By 1980, I decided it was time to explore the big-time world of magazine publishing. I realized that if I were to succeed as a professional underwater photographer, I must conquer not only the sea, but New York City, the center for the world magazine market.

22-23 A frozen barrel of fish entrals is used to attact a pack of reef and nurse sharks. At times food is used to help sharks lose their fear of humans and come within camera range.
Walkers Cay, Bahamas.
Depth: 16 meters.

23 bottom Ned Watson, age 13, enjoys the wonderful insanity of a pack of stingrays.
Stingray City, Bahamas.
Depth: 9 meters.

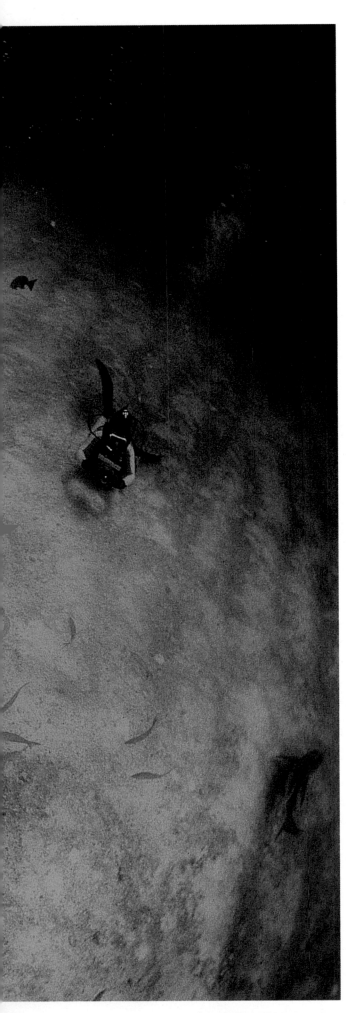

On a whim and with considerable chutzpah, I called *Life Magazine*. The picture editor, Mel Scott, agreed to give me a few minutes. With sweaty hands and cotton mouth, I took an all-too-short elevator ride to the 32nd floor of the Time-Life Building. I walked through a corridor lined with *Life* photographs, all with one common characteristic — "stopping power" — the kind of impact that prevents you from turning the page to see the next photo. This gallery of kings intimidated me, but Mel Scott quickly put me at ease. He complimented my macro photography and said he would keep my portfolio and show it around. Two weeks later Mel called to offer me a five-page spread in *Life*. My happiness was undescribable... I was going to be published in *Life Magazine*!

By this time my passion for underwater photography had taken over my life. I took a sabbatical from teaching. *Time-Life Publications* had just launched a new science magazine called *Discover*. On the strength of my first feature in *Life Magazine*, the editors at *Discover* were willing to let me fill an eight-page gallery section with photos on the Red Sea. Their instructions were simply to "shoot a smashing pictorial." *Life* gave me an assignment as well, to shoot a story on the schooling behavior of fish. And *National Geographic World* supplied the icing on the cake by giving me an assignment to shoot Israeli children enjoying the splendors of the Red Sea. I pinched myself and wondered if it was all a dream. I begged, borrowed, and stole in order to finance my return to the Middle East in the summer of 1980. At that time, the Sinai Desert, which bordered one of the richest areas of the Red Sea, was Israeli territory. I took up residence in Jerusalem and rented a tiny apartment in Sharm el-Sheikh. I dove into my work with a passion. Every day was an education.

The schooling story for *Life* offered an exciting challenge. The northern Red Sea was full of fish, many of which school. But capturing that schooling behavior on film was quite another matter. *Life* had impressed upon me that pretty pictures are great, but they had to "tell a story." I soon located a dazzling school of glassy sweepers. Every day I visited the school to observe their behavior. In the fading late afternoon light, a sizable grouper would suddenly appear, and without warning, dart into the school and grab a fish. The grouper's attack dramatically demonstrated the purpose of schooling. I needed to catch that moment on film when one fish was sacrificed for the survival of the rest. Finally, my patience paid off. The grouper darted into the school and missed its target. The sweepers began schooling in circles around their tormentor. The grouper was

utterly disconcerted by this turn of events, and I managed to get off a shot which epitomized schooling behavior. I now had the picture that would make the story complete! *Life* ran the feature under the headline, "Ex-teacher dives into fish schooling: Class Pictures." Becoming a full-time underwater photographer had its advantages and disadvantages. The positive part was being able to spend more time in the water than I had ever dreamed possible. The down side was the enormous expense of organizing my dive trips, usually to exotic, hard-to-reach locales. There are few full-time underwater photographers in the world. Such a low number is not due to lack of interest or desire. Financing the costs of these expeditions was prohibitive, yet I wanted to maintain the freedom of being a free-lance photographer. The sale of a magazine feature article to one magazine in the United States would barely cover the cost of the shoot. At this rate, I could be successful and in debt. I started to look toward Europe as a potential market for my work. European art directors would lay out my stories in unexpected ways that produced dazzling and dramatic results. I had found my solution to the crushing expenses of underwater photography... sell, sell, and sell. During this period I began working with an especially talented writer, diver, and scientist, Dr. Joseph Levine. Having received his

Ph.D. in marine biology from Harvard, Joe was that rare researcher who could write about science accurately but in a popular style that laypeople could understand. Together we could sit down, brainstorm, and argue our ideas through to completion about the subjects I was most interested in photographing. With this marriage of photographs and words, we became the darlings of science-oriented publications.

Underwater photojournalism, documenting a news story beneath the sea, is an exciting field that offers its own special challenges. During that time I begun diving at night: my favorite time to experience the mystery of the sea.

If I was offered the choice of doing one last dive, it would begin during the changeover time between day to night, extend throughout the night, and end with the transition from night to day. Changeover time, be it dawn or dusk, offers action with a capital "A". This is a dangerous time for the animals of the reef. Imagine having to hunt for a meal and being hunted at the same time. Being predator and the prey simultaneously is a stressful situation. You can sense the nervousness of the reef fish as they scurry between their daytime feeding grounds and their nocturnal shelters. Night sharpens the adventure and discovery of diving. Because most diving is done during the daylight hours, far less is known about the nocturnal sea and its creatures.

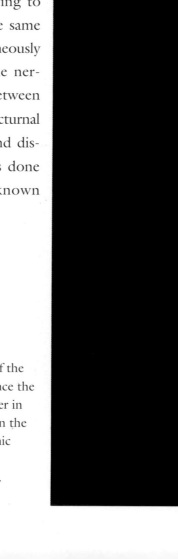

24-25 Paradise regained? Nearly a half century after atomic bomb tests turned Bikini Atoll into a nuclear graveyard, its tranquil beauty, both above and below sea level, have been reclaimed. Here Jeff Davis at 30 meters explores the deck of the U.S.S. *Saratoga*, once the largest aircraft carrier in the US fleet, sunk in the second of two atomic tests in 1946. Bikini Atoll, Pacific. Depth: 30 meters.

25 top The *Marex* salvage team raises an ancient cannon, part of the arsenal of the *Maravillas*, a ship of the Spanish fleet, which sank after straying onto the Little Bahama Bank on January 4, 1656. Little Bahama Bank, Bahamas.
Depth: 12 meters.

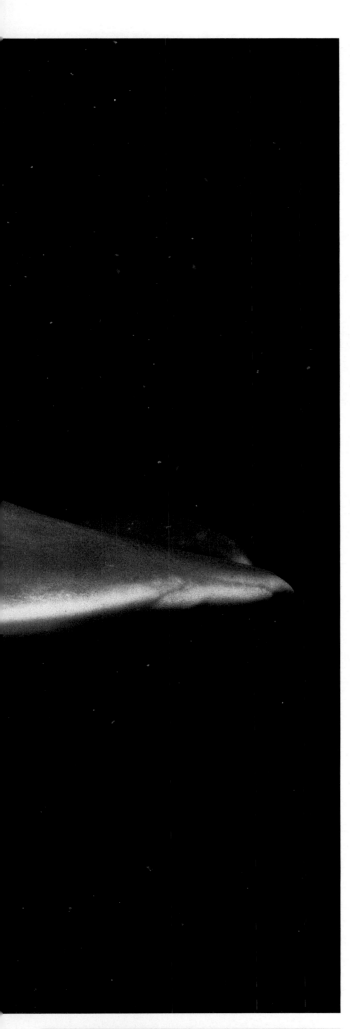

The cover of darkness offers both you and the marine community the opportunity to move about unnoticed. It also creates an unusual opportunity for intimacy with the creatures of the sea. During the day, I am lucky to get within a meter of a wary parrotfish. But at night I might be able to gently cradle a sleeping parrotfish in my hands or move a macro lens to within 3 centimeters of its eye.

If you only dove during the day, you would miss the greatest show on earth, or at least in the sea. Many performers only come out at night. Basket stars wave their weblike tendrils in the rich planktonic current. Sea urchins emerge from coral caves to hunt, and all manner of predatory beasts appear out of the blackness. People frequently ask me what I consider to be the best diving spots in the world (as I, as a novice diver, asked Jacques Cousteau). My easy, evasive answer is that they are all great, it simply depends on one's capacity to appreciate the varieties of marine environments.

While the cold, temperate oceans might take some acclimatizing, the life within these chilly waters constantly overwhelms me. In New England waters you may encounter the ugly but astounding Atlantic goosefish, with a mouth big enough to swallow, if not a goose, at least an entire seagull. In the North Pacific, there is no more impressive sight than a jetting and inking giant Pacific octopus. Southern seas are remarkable for their color and clarity. In the Bahamas, where you can see more than 60 meters in any direction, it is possible to work in open water with miles of ocean between you and the bottom. Sipadan Island, off the coast of Borneo, is a turtle-lover's paradise. I once watched a pair of turtles mate for over forty minutes in a shallow cave, until I had to return to the surface for air. The waters around Dangerous Reef, Southern Australia, offer the great opportunity to experience an encounter with the six-meter-long great white sharks.

If there is one ultimate dive site, I have yet to discover it, but that certainly does not mean I will stop trying!

I feel like my undersea journey is just beginning. The more I see, the more motivated I am to explore further. At one time, the number of dive sites around the world seemed infinite. Now, even the remotest islands are accessible to sports divers.

The ocean is far more fragile than we originally thought. Only we, the ruling and determining species, can decide if diving locations throughout the world will be preserved. As our understanding of our planet grows, we are faced with decisions about exploitation or conservation, enjoyment or abuse, that are truly choices of consequence.

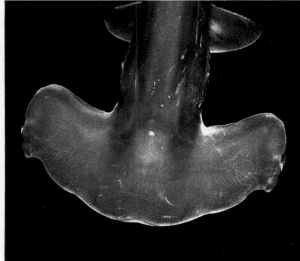

26-27 The head of the hammerhead shark (Sphyrna lewini) is one of the most effective instruments in the sea. The movement of this work of art on its muscular body makes you humble. Kanehoe Bay, Hawaii. Depth: 18 meters.

28-29 The pupil of this young hammerhead shark (Sphyrna lewini) is large, compensating for dim light conditions. Kanehoe Bay, Hawaii. Depth: 18 meters.

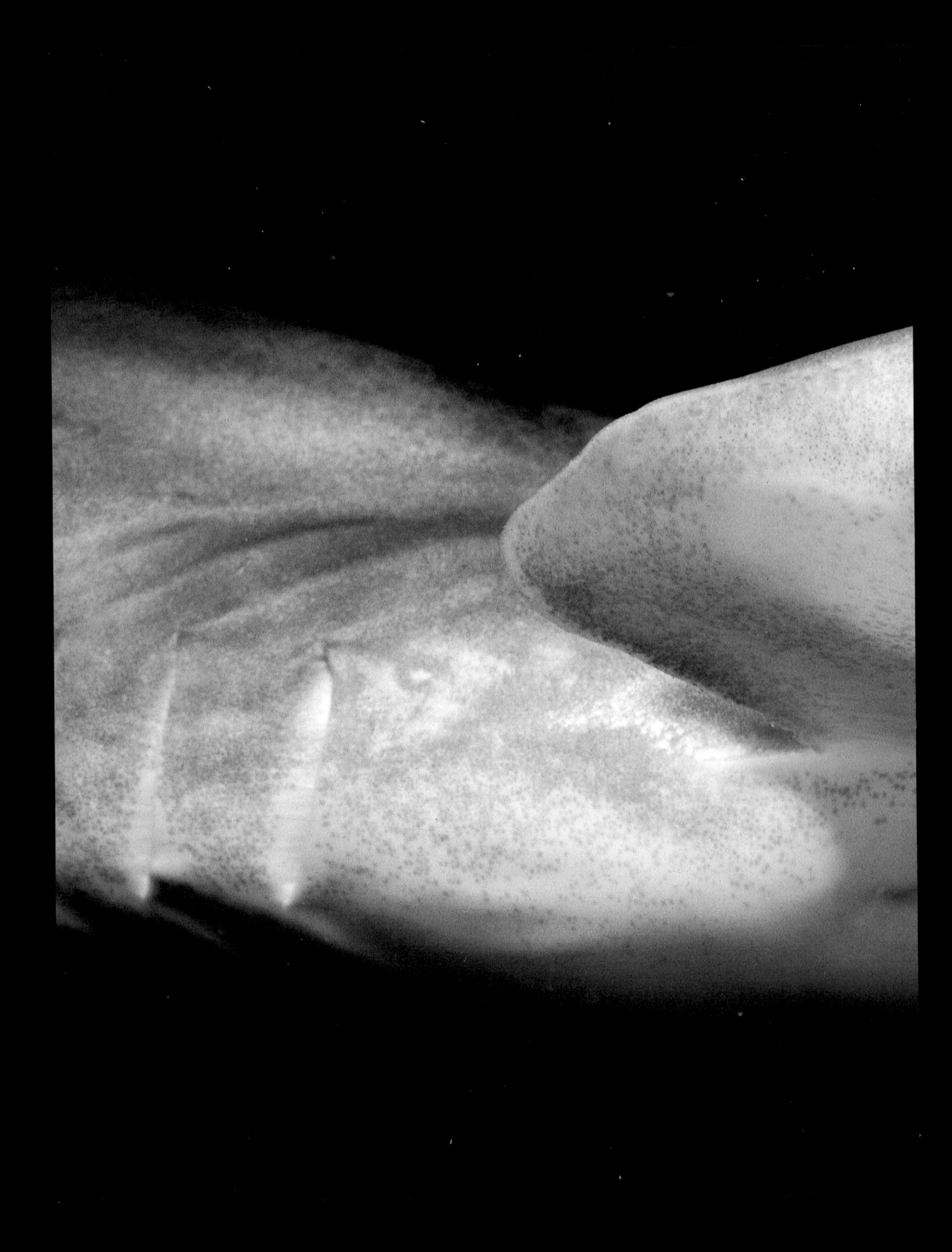

ATLANTIC

I fear, respect, and love diving in the Atlantic. It is the place of my birth — my entry into the world of the sea. It is where my underwater journey began. I had been a beachcomber since I was a child, walking the surf line at low tide and examining the creatures that the sea offered up. There were the horseshoe crabs — little changed from the time they first appeared in the oceans over 200 million years ago — which came out to lay their millions of eggs each Spring on the sandy beaches. Crabs scuttled everywhere, never reluctant to pinch my stubby toes with those powerful armored claws. I had never found more than the washed-up claw or molted shell of a Maine lobster along the shore, but I had heard that giant lobsters crawled all over ocean floor. But the greatest sea monsters of all, I knew, were sharks. Every summer, a fishing tournament would produce a record-sized shark. The local newspaper would run a picture of its carcass dangling from a pier hook, surrounded in death by jeering onlookers.

In 1973, the release of the hit movie *Jaws* reinforced in technicolor every shark tale of terror I had ever heard, and it spread fear of the Atlantic Ocean worldwide. No wonder I approached the Atlantic with apprehension. As a child in the port of Boston, I can remember hearing frightful stories of sailors who went down with their ships. Newspapers featured harrowing tales of people who lost life and limb to the fierce Atlantic.

Growing up, I had weathered the hurricanes and nor'easters that battered the New England coast. I had felt the pull of a strong spring tide when, standing knee-deep at the seashore, I imagined the watery forces that would drag me out to sea. All these memories flooded back to me when I started learning to free-dive in New England. On my first journey beneath the waves I was taken to a quiet, protected cove at a slack tide — a time when the tide stands still. My trusted friend Ken told me that the bottom was six meters below. Getting there was my goal. As I followed Ken's jump into the icy water, my apprehension was displaced by a curiosity about what lay hidden beneath the waves.

It was that curiosity that has kept me returning to the sea time and time again.. Coughing and sputtering after that first dive, I realized how much practice I needed in order to master free-diving. But gradually it happened. Eventually, I was able to explore these waters freely using only a snorkel. Moving without a cumbersome tank strapped to my back gave me a jubilant, light feeling, but eventually, the time limitation imposed by holding my breath made me switch to scuba.

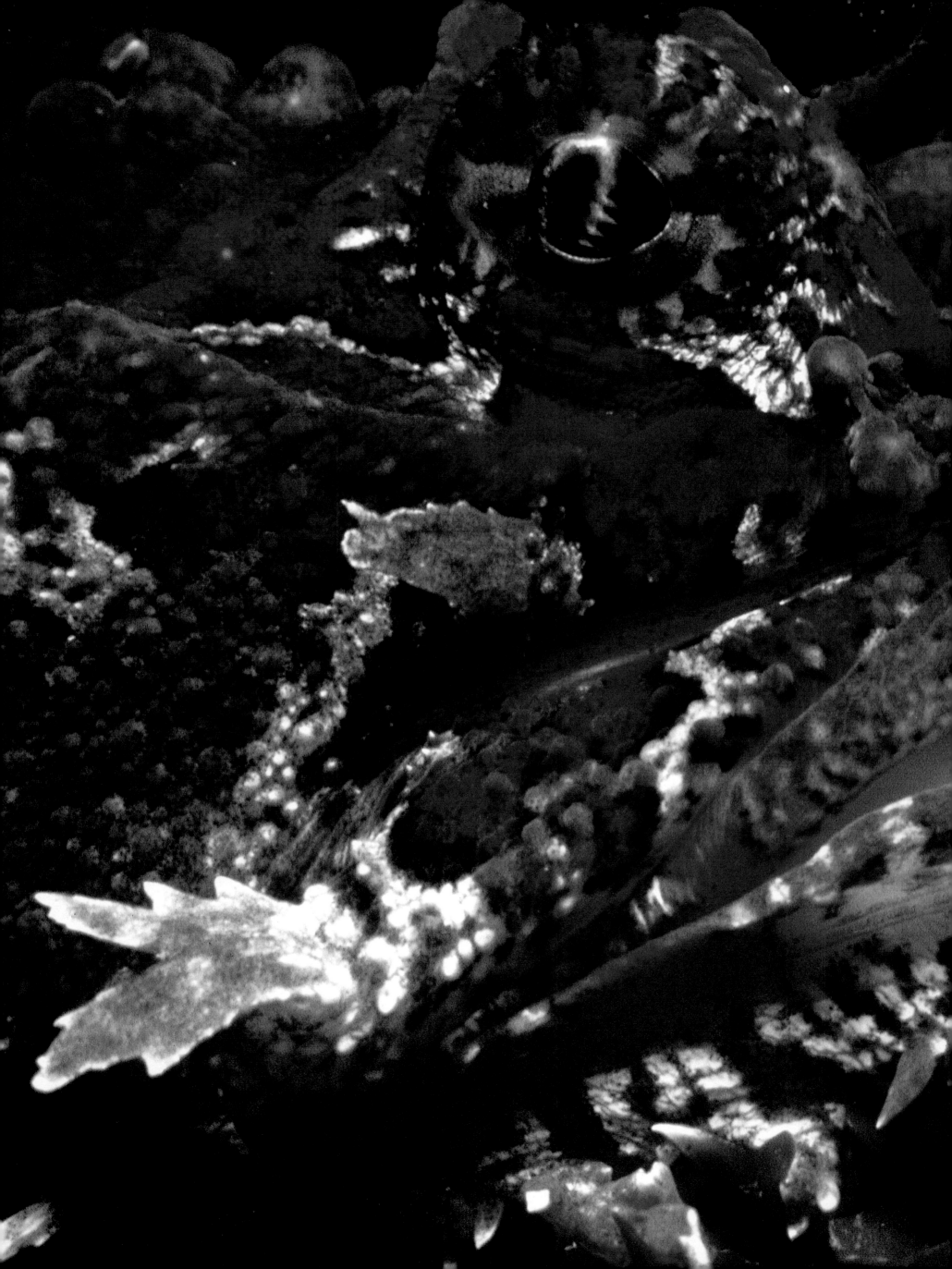

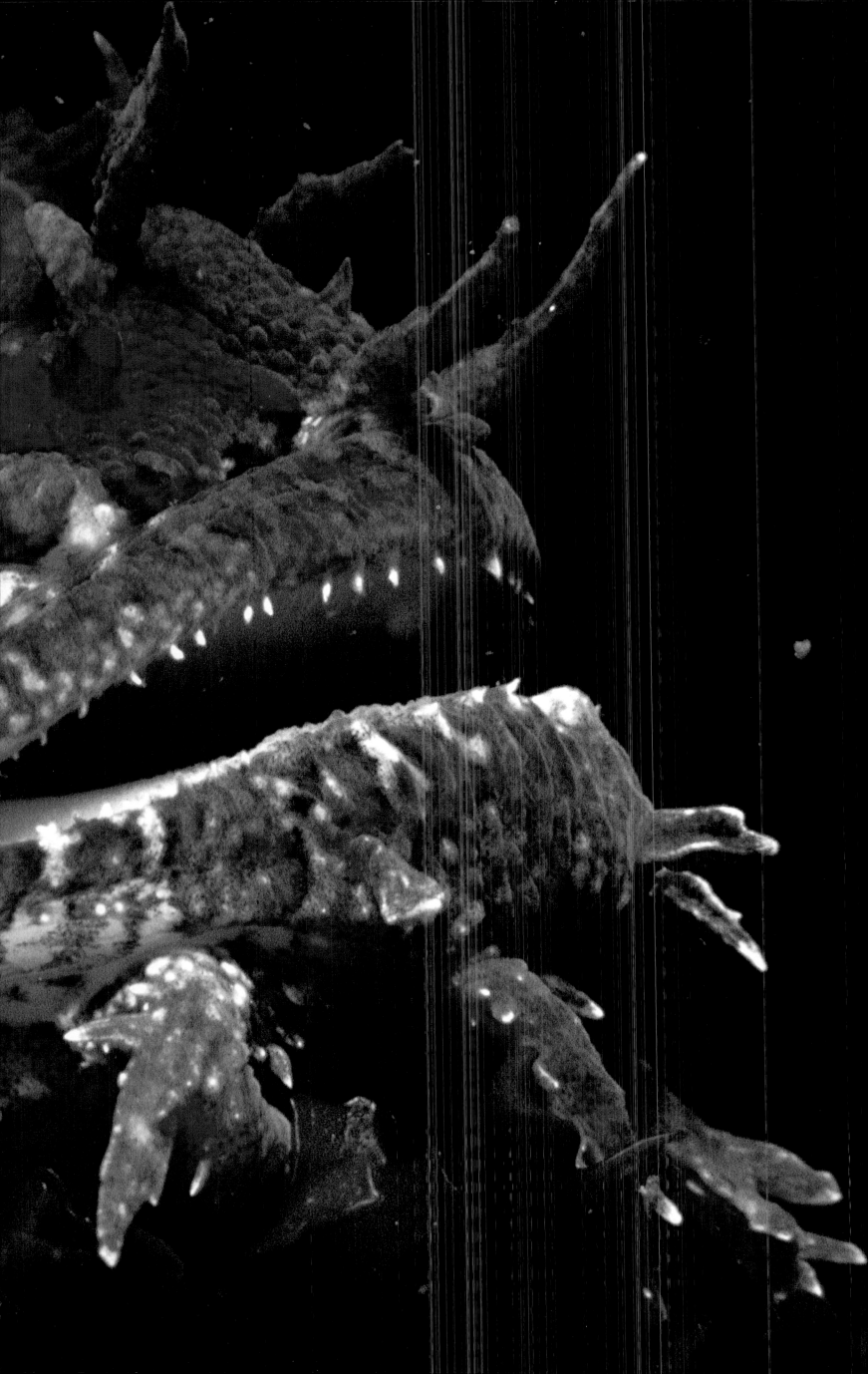

Even so, I still prefer free-diving. You experience the sea in a special way, almost as if you are part of it, and it gives your time with a scuba tank a whole different meaning.

Soon, I was traveling the entire Gulf of Maine from the great coastal plains of Cape Cod through Massachusetts, New Hampshire, and Maine into Canada. The population thins out as you head north from Boston.

By the time you reach the long coast of Maine, with its thousands of tiny coves and islands, you are usually alone on the shore, except for an occasional lobsterman repairing his traps. These fiercely-independent fishermen hunt long hours with wooden or wire traps that must be pulled, cleaned, baited, and dumped back into the sea each day.

Once, while trying to decide whether to dive a certain cove in Maine, I turned to a lobsterman suspiciously eyeing my dive gear. I asked him, "What's it like down there?" He answered with just a touch of sarcasm, "Don't know, don't live down there." In fact, he knew it better than I ever could. Having worked these waters a lifetime had imprinted a mental map detailing every nook and cranny of the ocean floor where he so carefully deposited his hundred-plus traps.

It was in Maine waters at a depth of only five meters that I first encountered a giant lobster.

Giant lobsters, any ones larger that thirteen centimeters from head to the beginning of the tail, are protected in this state alone. This one had to weigh fourteen kilos. It was trying to get at the bait inside a lobster trap, but it was almost larger than the trap. I wrestled it onto its back to find that it was a female with perhaps a hundred thousand tiny eggs glued to her tail. She would carry those eggs — berries, as the lobstermen called them — tucked under her tail for ten months until they hatched into tasty plankton. Despite her care, the odds of these babies surviving to dinner plate size were about one in 10,000. This lobster must have been very old. I was impressed. One of the delights in working the Atlantic has always been the bottom fish, or groundfish, as they are called by the fishing industry because that's where they catch them. Bottom fish seem to have a few things in common: they are usually "ugly" by our human standards of beauty. Some have bizarre, fleshy tabs waving about their faces like the seaweed on the ocean floor. Their dull, earthy colors also camouflage them against the sea bed. Many bottom fish have gigantic mouths that gulp down whole fish. They actually look as if they are inhaling their prey.

The flounder, despite its tiny mouth, is another ludicrous-looking bottom dweller. Both its eyes are on one side of its head!

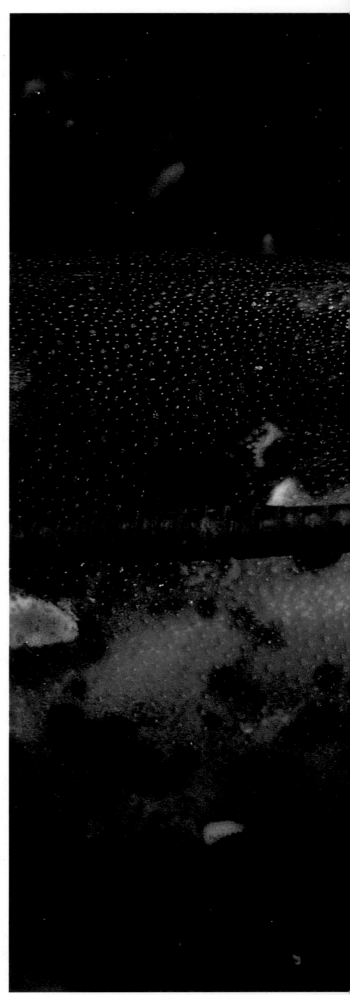

34 bottom At the end of summer I came upon a mated pair of Jonah crabs *(Cancer borealis)*. The female had just molted, the male crab was above, protecting her. Hodgkins Cove, Cape Ann, Massachusetts. Depth: 5 meters.

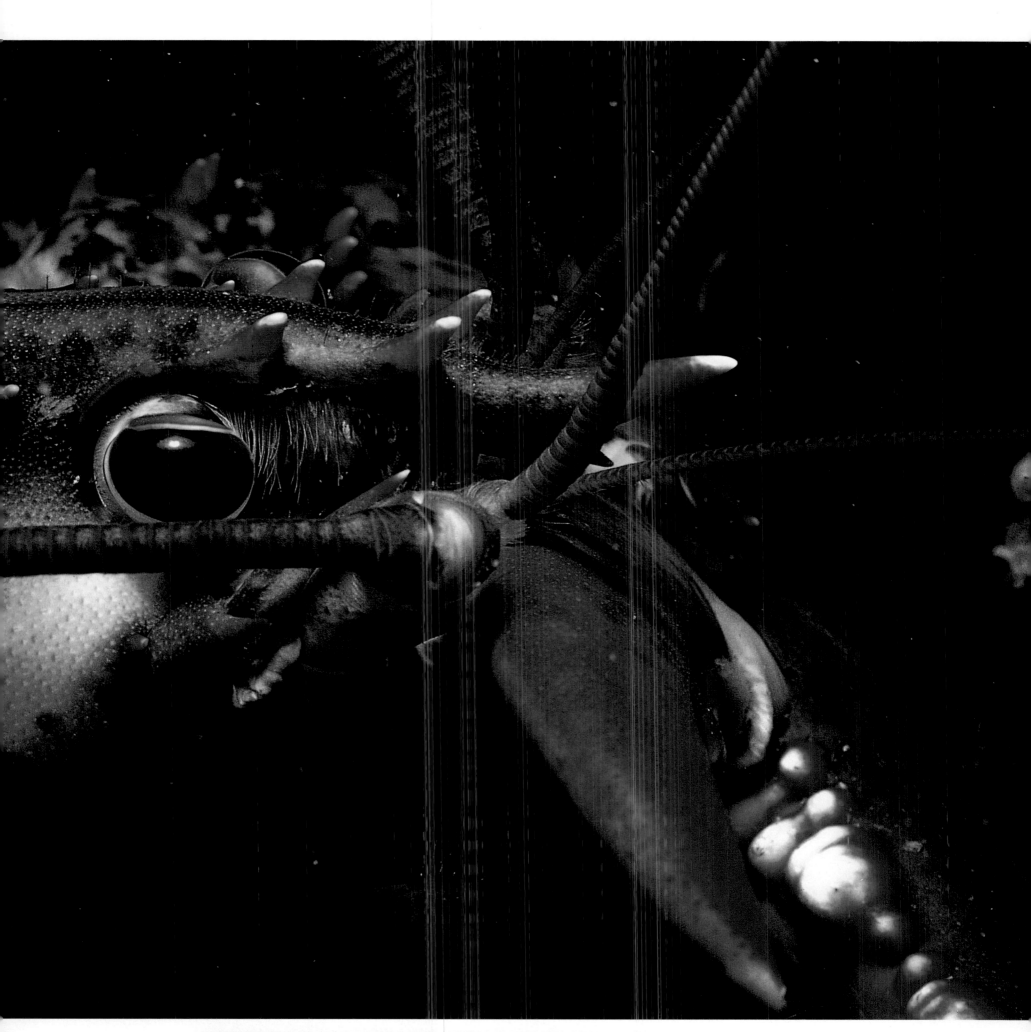

34-35 Formidable in its
armor and with sharp
spines bristling from head
and claws, the North
American lobster
(*Homarus americanus*)
can repel many potential
predators.
Cape Ann, Massachusetts.
Depth: 6 meters.

35 bottom Gaping
wide, horse mussels
(*Modiolus sp.*) use hair-like
cilia to propel a current
of water over their
internal gills, which are
modified to filter
planktonic food. The
shells of the mussels are
encrusted with pink
coralline algae. Although
the purple seastar is this
mollusk's natural
predator, the immature
individual seen here is
too small to be a threat.
Cape Ann, Massachusetts.
Depth: 22 meters.

36 top Sea Butterfly or pteropod, *Clione limacina*, normally lives in deep water well offshore but at night comes to the surface to feed. It is occasionally seen in coastal waters of the Atlantic in the spring. This individual, only two centimeters long, was photographed just days after the great blizzard of February 1978, which probably blew it in towards shore. They are eaten by baleen whales. Folly Cove, Gloucester, Massachusetts. Depth: 3 meters.

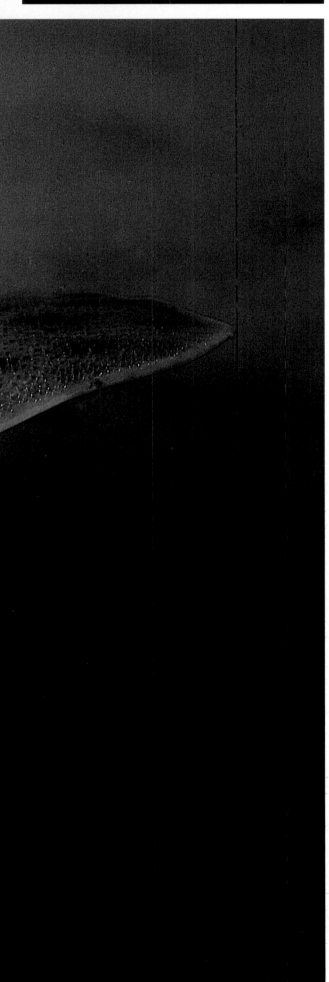

36-37 Put to flight, this young skate *(Raja sp.)* was surprised at dawn by my companion's underwater lamp. These fish enter sandy coves in spring and summer to reproduce.
Cape Ann, Massachusetts.
Depth: 2 meters.

37 top The moment captured here was not fully appreciated until this photograph was processed, enlarged and contemplated.
If you look closely, you will see two planktonic amphipods curled about the body of a third, possibly for the purpose of reproduction. These animals are less than one centimeter long.
Cape Ann, Massachusetts.
Depth: 4 meters.

When the flounder is still quite young, one eye migrates across the skull until it joins the other eye on the opposite side of its head.
This adaptation allows it to lie flat against the ocean floor while keeping a look-out for prey with both eyes. Imagine how we would look if we had both our eyes on one side of our face! Picasso must have been inspired by flounders.
Except for the fishes, much of the marine life in the Atlantic is sessile or slow-moving.
Sea anemones do not run from your lens, but they will retract their tentacles into a jelly-like blob unless you approach them carefully.
Tiny nudibranchs with fashionable colors lay hidden like treasures in this dark, gloomy sea. Only the power of the light of a diver's lamp reveals their true colors.
The weak do not survive here for long.
Cannibalistic moon snails drill holes into the shells of their relatives, like mussels, clams, and oysters. If a hermit crab covets a bigger snail shell for its new home, it will most likely have to fight another hermit for it or devour the snail within. A female lobster can only mate just after she has just molted her corset-like shell when she is defenseless. But her mate will embrace her fragile body and act as her bodyguard until her new shell hardens.
Quantity is the key to survival in the Atlantic. While the variety of life is limited here, offer-ing far fewer species than in the tropics, the animals that gain a niche in this hard, unfor-giving environment occur in great abundance, or used to. Gone are the great schools of hali-but as well as those of cod and flounder, and a host of other groundfish.
They have been fished out to the point of col-lapse of the species. The slow response by government and fishermen to the warning signs of decreasing catches allowed this calam-ity to happen, resulting in the devastation of this great marine food basket.
Once the rich fishing and breeding grounds of Georges Bank seemed inexhaustable, but that was before technology caught up with the fish, pursuing them with sonar and spotter planes until they had no place left to hide. Ghost nets drifting across miles of sea caught fish that would never be harvested by man.
In the twenty-five short years that I have been diving the Atlantic, the change in the quantity and kinds of species has been striking.
Cod, haddock, and flounder have been replaced by underutilized species like skate, dogfish, and squid.
We seem to have proved we are not capable of managing our marine resources.
Only man seems to be able to alter this marine environment in ways that marine organisms find impossible to withstand.

38-39 The cod fish *(Gadus morhua)* - synonymous with the Atlantic Ocean and once numerous beyond fantasy is now endangered as is the rest of this special food basket - New Englands' once "rich Atlantic."
Cathedral Rocks, Massachusetts.
Depth: 27 meters.

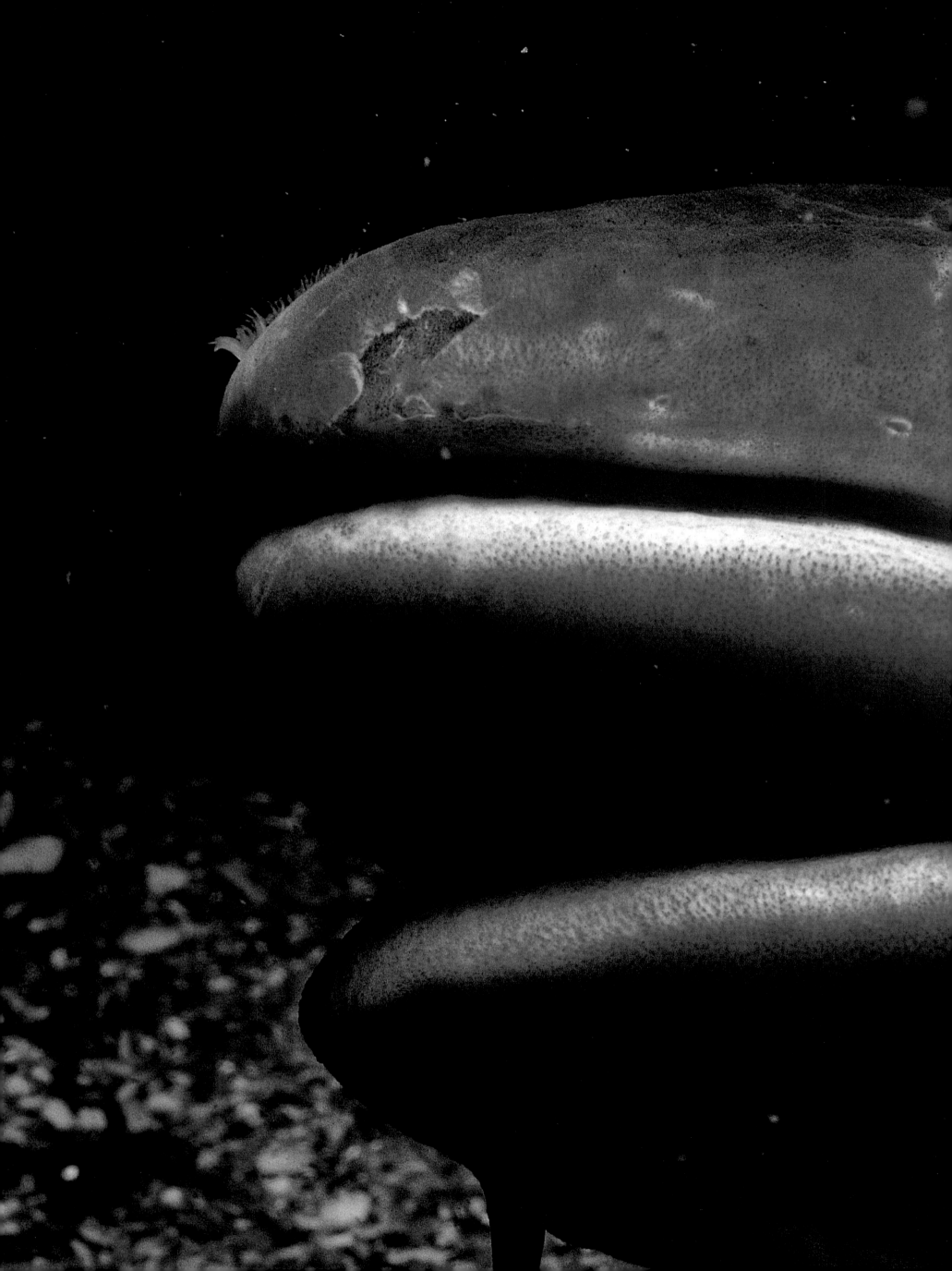

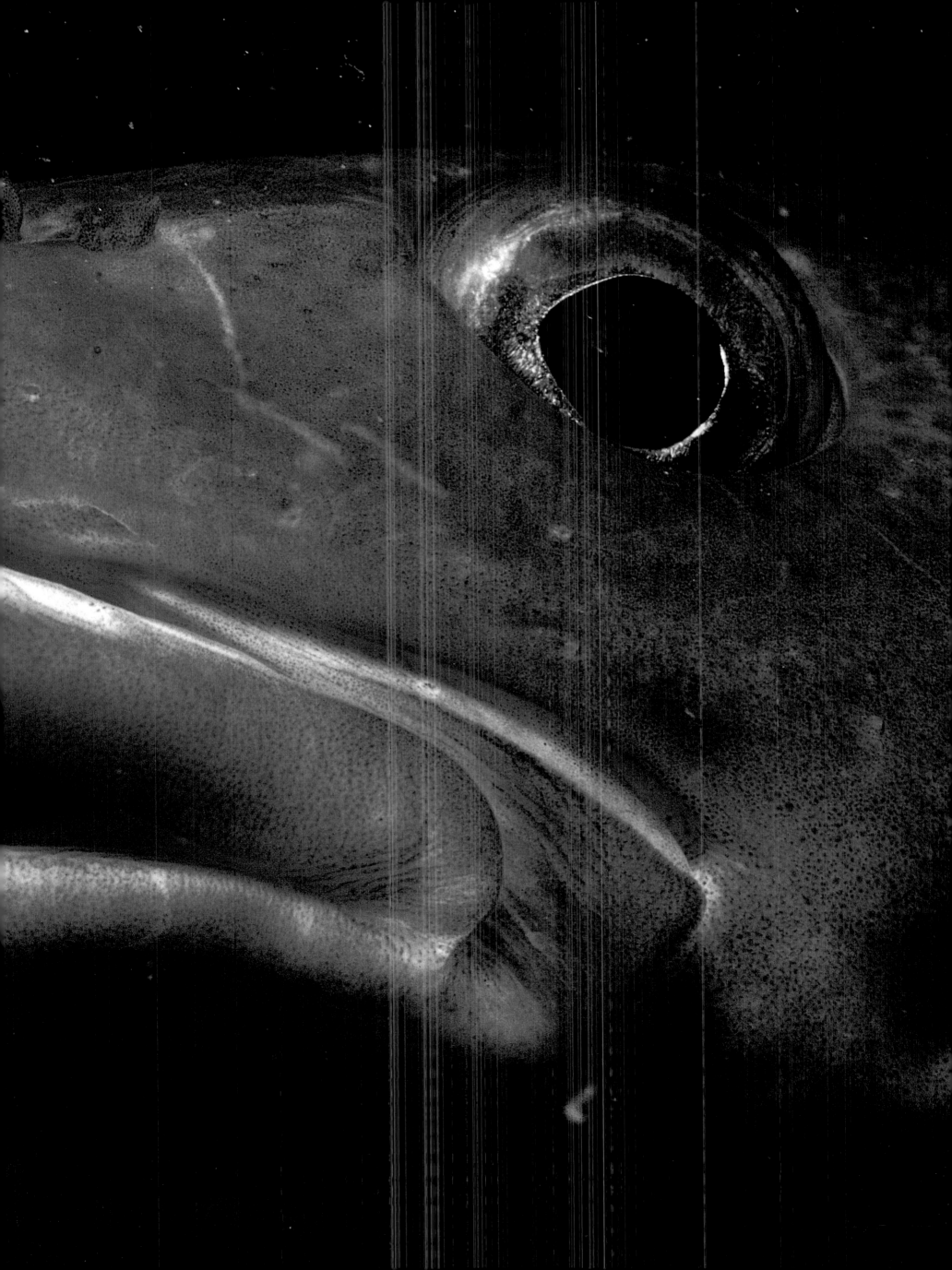

In 1993, the French nature magazine, *Terre Sauvage*, gave me the opportunity to photograph the other side of the Atlantic. I traveled to Glenant Island, an offshore island in southern Brittany. Situated in an area washed by fertile ocean currents, it promised a rich cross-section of temperate coastal life.

I expected to find some animals similar to those I knew from New England, but there were many new friends to meet in these chilly waters. Accompanied by my assistant, Jerome Delafosse, a budding young French filmmaker from this region, we jumped into green, murky waters just like on the other side of the Atlantic. The sediment and plankton suspended in the water threatened to make my photographs appear as if it were snowing underwater. As we neared the bottom at fifteen meters, the visibility cleared a bit. We crawled over rocks draped in pink coralline algae, inspecting the substrate for life.

There were creatures everywhere. Hermit crabs picked at the interstitial sand scavenging for tiny organisms and cleaning the ocean bottom at the same time. Empty mussel shells revealed the presence of an active starfish population. Marauding starfish left a trail of empty bivalve shells in their wake. Suddenly, we came upon a pack of the culprits. It was *Asterias*

vulgaris, the same purple starfish that I knew from New England waters. A giant mussel was enveloped by seven separate animals, their suction-cup arms wrapped in a stranglehold around the mollusk's shell. Soon the mussel would tire, its shells would part, and the starfish would thrust their stomachs into the unfortunate animal to digest it inside its own shell. Moving just offshore of the coast, we clung to the interface of rock ledge and sandy coralline algae bottom. Where these two zones met was a richer hunting ground and improved our chances of finding subjects to photograph. Suddenly Jerome waved at me frantically to join him. I could see as I approached what had excited him so — it was a goosefish.

The French call it *lotte* and charge outrageous prices for a sliver of this fish on a plate. If they could see the whole animal, diners might not be so enthusiastic. Its look combines ugly, bizarre, and humor all in one. The goosefish usually lies camouflaged on a mud or sand bottom, luring unsuspecting fish with "a fishing

40 The astiral symmetry of this sea star *(Asterias sp.)* is accentuated as it transits a blade of kelp. The spare simplicity of this moment seems to capture the essence of all plant and animal life. Glenant Island, Brittany. Depth: 10 meters.

41 The tubed feet of the starfish arm reach out to "see" its environment. By touching it is able to taste. Glenant Island, Brittany. Depth: 7 meters.

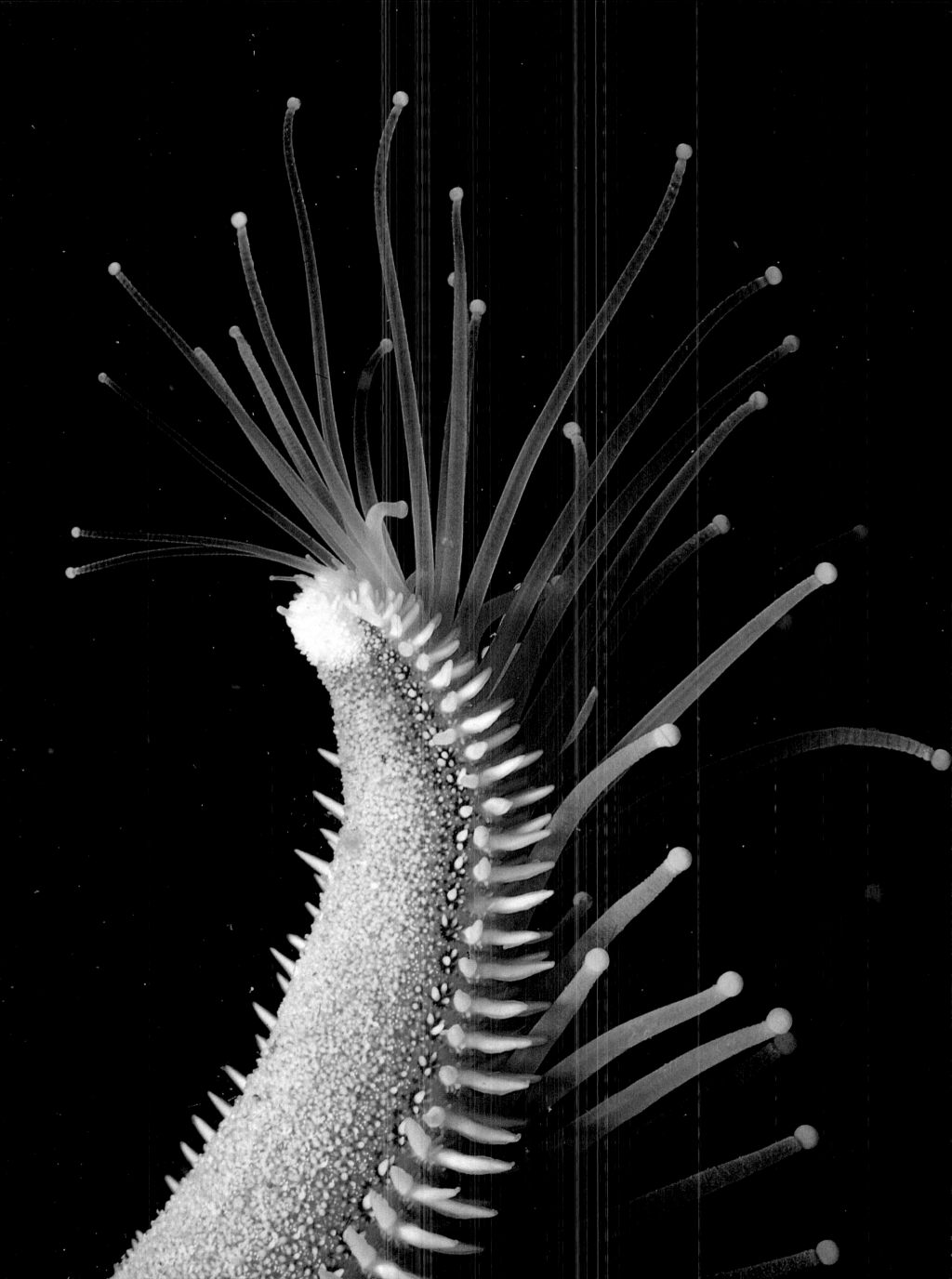

42-43 Ugly is beautiful.
The goosefish *(Lophius
sp.)* lies camouflaged on
the sandy bottom to lure
unsuspecting fish with the
angling appendage
carried on its upper lip.
Pocketbook mouth and
sharp conical teeth make
its victims disappear with
surprising ease.
Glenant Island, Brittany.
Depth: 10 meters.

44-45 A mass of tangled
arms prevents you from
seeing the prize
underneath - a tasty
mussel.
Glenant Island, Brittany.
Depth: 8 meters.

pole" mounted on its upper lip. Its pocketbook mouth and sharp conical teeth make its victims disappear with surprising ease.

One afternoon when I was getting low on air and beginning to think about the warm, after-dive shower, I came upon a fish that made me forget terrestrial pleasures. Lying on a moon-like surface twelve meters deep was an absurd-looking fish that we call a sea robin in America and *rouget grondin* in France. Its fleshy, crimson mouth reminded me, perversely, of Julia Roberts' sensuous lips. Its birdlike pectoral fins had almost evolved into legs. In fact, it crawls along the ocean floor using those sensitive fins to detect prey. The sea robin brought thoughts of a fish changing into an amphibian so many eons ago. It eyed me suspiciously with its over-sized orbs, as I focused on the color and texture of this marine "work of art.". It tolerated my intrusion only for a few flashes of my strobe, but it left me with one of my favorite fish faces.

As I tore down my gear and prepared it for shipping, I reflected on the similarities and differences between the two sides of the Atlantic. Both were rugged marine environments that brought out aspects of "gorilla diving." Their beauty was sly and subtle, painted in shades of mauve and earth tones. You had to work hard to get them to reveal themselves.

It took powers of concentration to see what was here, but it was worth the effort.

It seemed so healthy and rich, but overfishing has threatened the ecological balance on both sides of the Atlantic. Too many divers pass up the spendors of the North Atlantic in favor of the more glamorous tropics. I think that is their mistake... and ultimately their loss. Each different marine environment I see helps me to better understand the sea around us, and the Atlantic is an important and wonderful piece of that sea.

46 The dorsal fin of the Saint Peter fish *(Zeus faber)* displays its long, dramatic dorsal fin - possibly an attempt to make it appear larger to a potential predator. Glenant Island, Brittany. Depth: 8 meters.

47 This Saint Peter fish *(Zeus faber)* displays a bright yellow coloration making identification easy - perhaps by a prospective mate too. The language of color is still a mystery in marine science. Glenant Island, Brittany. Depth: 8 meters.

48-49 This sea robin *(Trigloporus lastoviza)* took first prize in my fish face competition on this side of the Atlantic. Watching it cross the bottom on pectoral fins that seemed to have evolved into legs brought thoughts of the first amphibian that crawled from the sea. Glenant Island, Brittany. Depth: 8 meters.

The Pacific Ocean, West Coast, USA

50-51 With their "shark grace" blues *(Prionace glauca)* are sexy in the water. San Clemente Island, California. Depth: 5 meters.

n 1979, I was still teaching science to 13-year-old children in Cambridge, Massachusetts. This allowed me to devote my summer exclusively to underwater photography. I had just begun work on my first book, *Beneath Cold Seas*, and I was on my way to the cool, temperate waters off the coast of California. My plan was to spend the summer working my way north along a thousand miles of Pacific coast into British Columbia to take an intimate look at this much-talked about marine environment.

That same year, I headed north after spending a month exploring the Channel Islands aboard Roy Hauser's live-aboard *Truth*. As I began exploring areas north of Point Conception, the water got progressively colder and the marine life changed accordingly. In these dark waters, many of the invertebrate animals sported bright, blazing colors. For certain poisonous nudibranchs and sea anemones, this coloration serves as a warning to first-time predators. Eventually I reached Vancouver Island in British Columbia, Canada. The previous winter I had picked up a bulky wetsuit designed for these frigid waters. Getting into and moving in it was nearly impossible, but I was as warm as a toast. My dives have always been long by sport-diving standards, and in these waters I was spending upwards of two hours in less than six meters of water. The visibility often dropped to three meters, making wide-

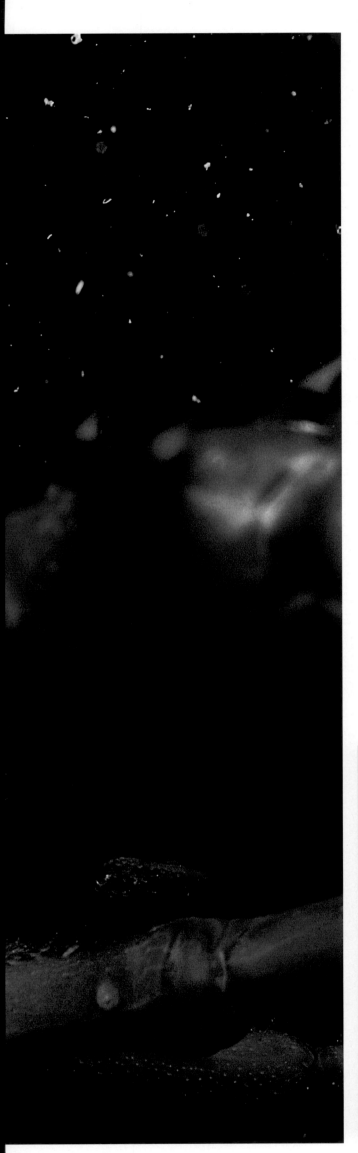

angle photographs almost impossible. But the macro life was superb. I was content. There was one animal that I had specifically come to meet that I had not seen yet — the giant Pacific octopus *(Octopus dofleini)*, that grows to legendary size here in the North Pacific Ocean. These shy and reclusive giants can weigh over 50 kilos, with arms that can easily envelope a diver. On the last day of my trip I was rewarded. At 15 meters below the first thermocline the water became crystal clear. As my dive light illuminated a cave, I saw a thick tentacle with double rows of suckers. Holy moon snails! I had found my first giant octopus! Living up to its reputation for seclusion, it refused to make an appearance. Three more dives to the den proved equally futile. Yes, I missed my giant octopus shot, but I would get another chance — 15 years later.

54-55 Like peppermint sticks, the brightly banded tentacles of the Christmas anemone *(Tealia crassicornis)* are arranged in concentric circles. This individual has drawn the first circle of tentacles to its mouth in what may be feeding behavior.
Monterey, California.
Depth: 23 meters.

56-57 Sharp spines keep predators at bay as these purple sea urchins *(Strongylocentrotus purpuratus)* creep imperceptibly along the bottom, feeding on algae and other organic matter.
Santa Rosa Island, California.
Depth: 7 meters.

52-53 The head of this spiny lobster *(Panulirus interruptus)* is equipped with sharp projections that are a formidable defense. Southern California.
Depth: 10 meters, at night.

53 Night feeders, strawberry anemones *(Corynacyis californica)* hungrily sweep the water with their tentacles. Santa Catalina Island, California.
Depth: 10 meters, at night.

In September, 1994, I packed up my aging dry suit to take another shot at the giant octopus that is unique to the chilly waters of the Pacific Northwest. This time, I teamed up with zoologist Jim Cosgrove of the Royal British Columbia Museum in Victoria, whom some people refer to as "Mr. Octopus." After working with him, I did, too. The waters that are the octopus' home are notorious for poor visibility because of the high plankton content. We had purposely chosen the month of September for our expedition, expecting that cooling temperatures and increasing rainfall would improve the water clarity. We guessed wrong. For a full week before my arrival, the sun nourished a healthy bloom of plankton. But Jim had been studying several individuals in the area, so he knew where to find them. As we descended along a rock wall, Jim peered into several small crevices in the cliff face looking for possible octopus hideouts. On the sea floor 23 meters below, he pointed toward a telling clue next to a cluster of boulders, as distinct as the X on a treasure map. A pile of empty scallop shells marked the entrance to an octopus den. Aiming a dive light into the cave, we saw two shiny black slits reflecting back the light. Our quarry was at home. Jim squirted a chemical into the den — a respiratory irritant designed to annoy

but not to harm the animal. Within seconds, a 30-kilos octopus gingerly emerged. When it spied Jim and me, it blanched white and dug its arms into the sand, prepared for a fight. We didn't approach. Jim wanted to let the octopus relax a bit first. Gradually the giant's curiosity got the better of it. It sidled over to Jim. It tentatively touched his mask with the end of one arm. Chemoreceptors on its sucker discs registered Jim's unusual texture and taste. It brought up more arms to investigate. It fingered his drysuit, mask, and tank as a rainbow of colors rippled across its skin. During the next week I would watch in amazement as octopuses parachuted silently down on unsuspecting prey and enveloped them with a net of arms, as one devoured a giant cabezon fish, and another, when cornered, squirted what seemed like gallons of ink. I was awed by these much maligned "monsters" that allowed us to handle them without ever trying to harm us. Their intelligent eyes, their graceful dancing movements, their shy, reclusive temperament, all touched my soul.

58 This giant octopus *(Octopus dofleini)* inked what seemed to be about twenty liters of ink in the water to camouflage its retreat and produce a ghost image, which some octopusologists think is another usage of inking. Victoria, British Columbia. Depth: 15 meters.

59 A full grown giant octopus *(Octopus dofleini)*, like the one pictured on these pages, can have an arm span in excess of 3 meters and weigh upwards of 50 kilos, and it can reach this size in less than 5 years! Victoria, British Columbia. Depth: 18 meters.

60-61 An octopus depends on its exceptionally acute senses of vision and smell to track its prey. An octopus' pupil is horizontal, rather than round like ours, but in other ways our eyes are very similar. Both human eyes and octopus eyes have eyelids, irises, crystalline lenses, and retinas. Victoria, British Columbia. Depth 13 meters.

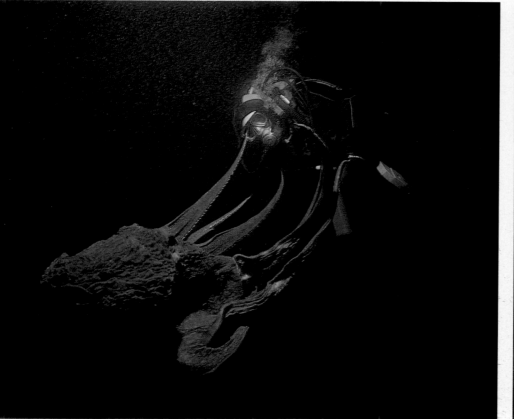

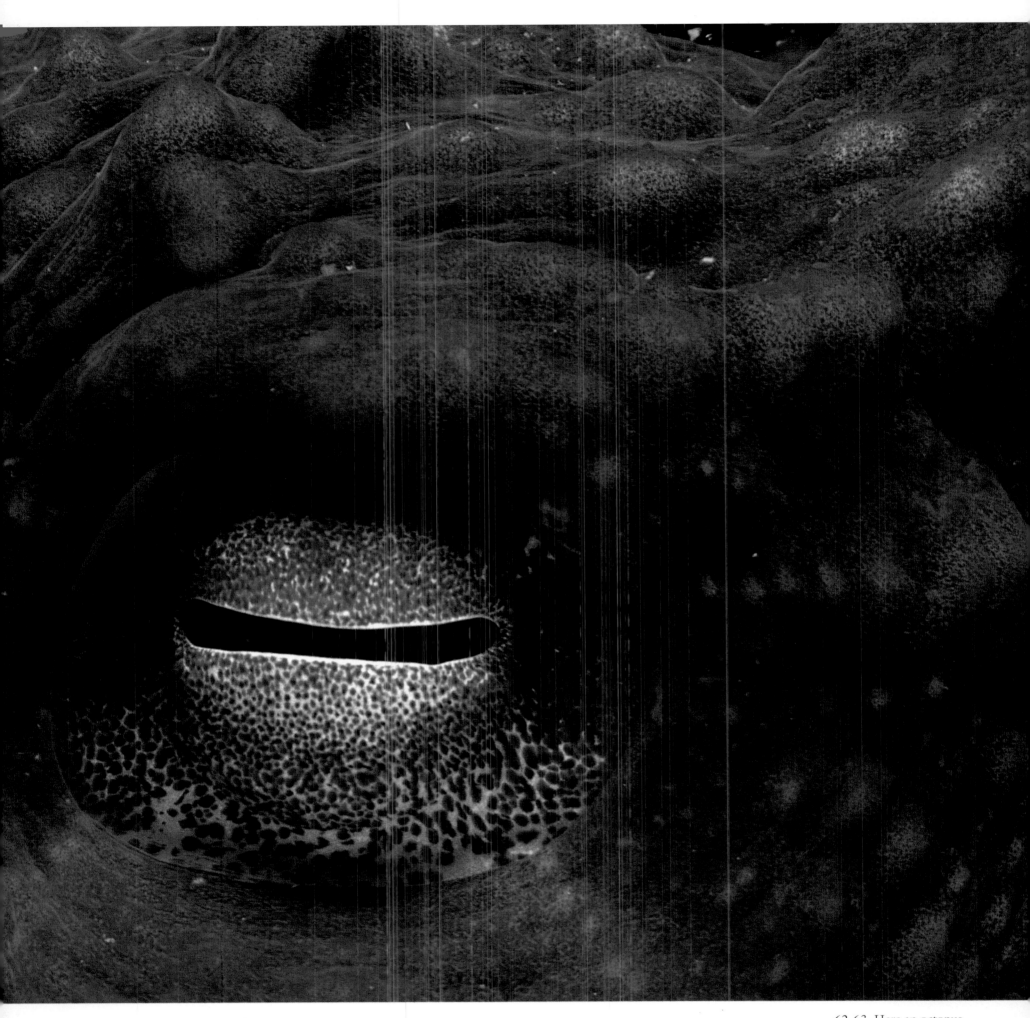

62-63 Here an octopus has located prey, a cabezon *(Scorpaenichthys marmoratus)* hiding under rocks, by poking its arms inside cracks and crevices. Sensory organs - chemoreceptors on the rims of the suckers can smell and taste. Victoria, British Columbia. Depth: 12 meters.

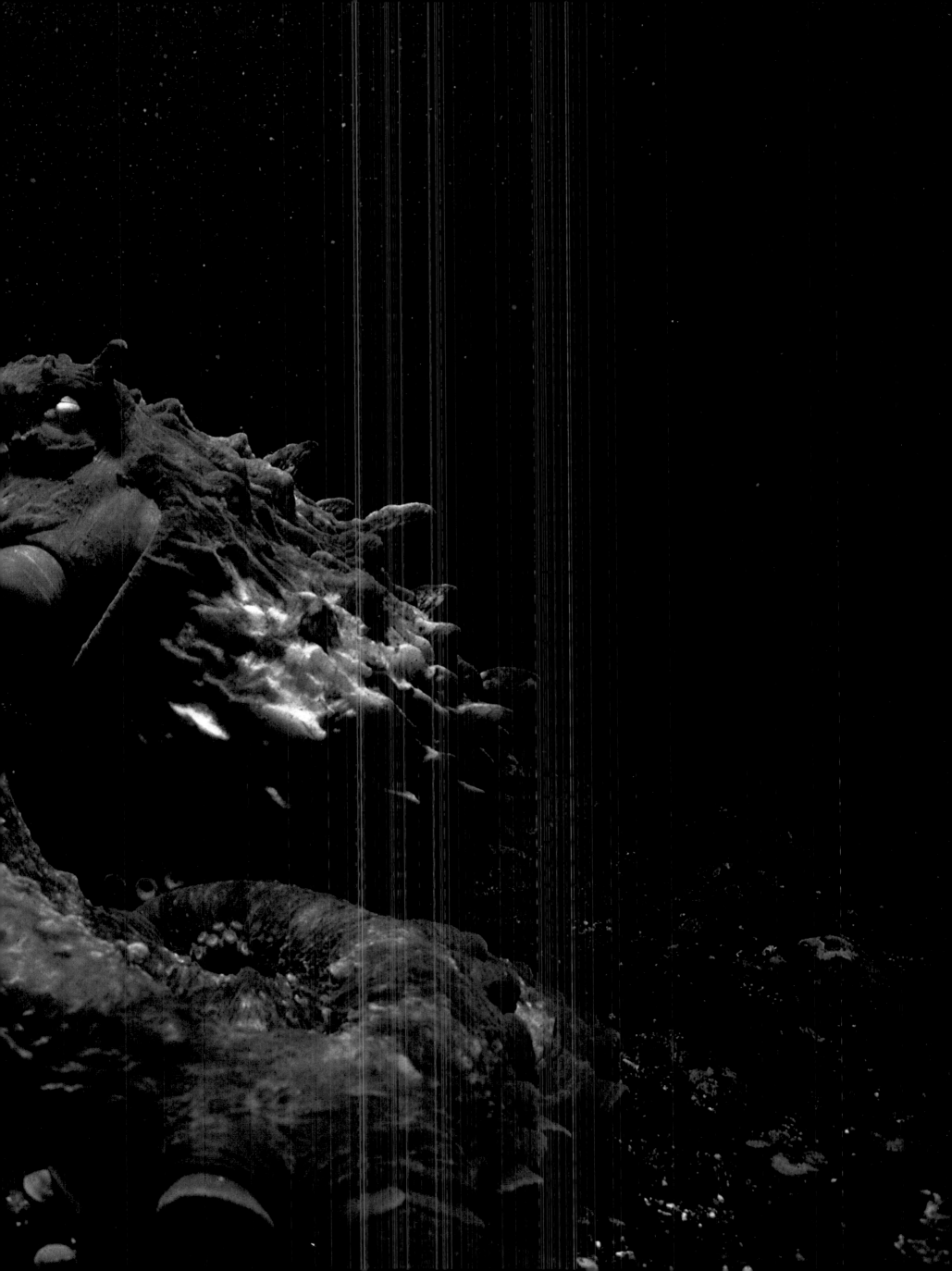

izarre marine animals seem to be present in greater abundance in most cold and temperate seas, so I was eager to find out what the California currents had to offer. The giant kelp forests are the structural foundation from which much of the marine life in southern California blossoms. From the base, or holdfast of this, the fastest-growing plant on earth, *Macrocystis* , reaches upwards of 40 meters toward the surface. On a good day, the plant may grow 60 centimeters; on a great day, it may sprout a meter!

From my first dive among the "green monsters," I felt like I had entered Rudyard Kipling's *Jungle Book*, with many and varied beasts hiding among the dark fronds of kelp.

At one moment I was buzzed by inquisitive sea lions that seemed to invite me to play, but I suspect they were probably just making fun of my awkwardness. A moment later, I detected the shadow of a cruising blue shark that had come to investigate me. It wore a "I'm always ready for a meal" grin on its face.

The larger predators lost my attention when my eye caught the impossible electric purple of a Spanish shawl nudibranch, its bright orange cerata whipping around in the surge.

Starfish in vivid, varied colors carpeted patches of the ocean floor beneath the kelp forest.

Large congregations of fish and invertebrates live in close proximity to the anchoring holdfast of a gigantic kelp plant.

Like Jack in the Beanstalk, I followed the stem of the plant toward the surface, finding a treasure of small marine life feeding on its nourishing pulp. Golf-ball-sized moon snails crawled along the plant's outer surface, scraping off an ever-so-thin layer of cells. Each snail was digesting its own distinct path up and down the length of the plant.

In mid-water, surrounding the kelp were schools of white-spotted *señorita* fish. They darted about individually but were quick to regroup into schooling formation at the first sign of a predator. On sunny days (and California is famous for them), the way the sunlight dappled off the kelp fronds in shimmering waves was humbling and spiritual.

California grey whales and their relatives were once commonplace here, but many species of these marine mammals are at the edge of extinction.

64-65 Stretching to over 40 meter in lenght, giant kelp *(Macrocystis pirifera)* create roomy undersea forests off the Pacific coast of North America. San Clemente Island, California. Depth: 17 meters.

66-67 Sheepheads
(Archosargus probatocephalus) and
kelp fish *(Gibbonsia sp.)*
part of a wide and diverse
kelp forest community.
Santa Catalina Island,
California.
Depth 10 meters.

66 bottom The electric
orange garibaldi
(Hypsypops rubicundus)
has enjoyed protection
from collectors and spear
fishermen since it became
California's state marine
fish.
Santa Catalina Island,
California.
Depth: 5 meters.

World-famous underwater cinematographer Howard Hall, the "king of the kelp forest," once photographed gray whales feeding on the kelp. As he was finishing his shoot, rising through the green, dreamy forest, Howard happened to surface right next to a California gray whale, its open mouth overflowing with kelp. Startled, Howard nonetheless had the presence of mind to shoot an award-winning photograph of the encounter. For those like Howard who spend most of their waking hours underwater, serendipitous moments like these are their reward for thousands of hours of patience.

In 1992, I joined Howard in the California kelp forest to shoot still photographs while he shot a documentary on Stan Waterman, *The Man Who Loves Sharks*. Stan's friend, author Peter Benchley of *Jaws* fame, was along for the ride and the laughs that Stan was always quick to provide. Back in the 60's, Waterman was the first *Homo sapiens* to film a great white shark in the epic film, *Blue Water, White Death*. Thirty years later, we were anchored off San Clemente Island, chumming with mackerel, trying to attract blue sharks. The belief was that the once-abundant blues had been decimated by indiscriminate gill netting. We had come to California to dramatize this ecological catastrophe — but it didn't work out that way. Within two hours of our arrival, the water was boiling with hungry, skit-

tish blue sharks, streaking through the water like slender, graceful torpedoes. As the shark cage was being lowered, we saw more than a hundred blues milling about. We immediately realized that the reports of their demise had been greatly exaggerated. The film producer was furious that so many blues in the frame would do nothing to reinforce the imminent extinction theme of his production. But Stan and Howard were having the time of their lives. Howard was clothed in his chain mail, anti-shark suit which had proved effective against small-mouthed sharks like these ocean-going blues. In a parody of feeding the sharks doggy bones (actually, tasty mackerel), Howard and the blues wrestled fish for Stan's camera. Protected only by a wetsuit, Stan nonchalantly moved among the blues looking for good camera angles and photogenic action, unperturbed by the sharks that seemed almost magnetically attracted to the dome of his movie camera housing. Perhaps his casual attitude had something to do with his safety diver, Bob Cranston. Clothed in full chain mail, Bob was knocking home runs and triples with a sawed-off baseball bat, batting curious sharks away from Stan with his home-made shark stick. It was really wonderful to watch Stan Waterman at work — this man of 70 years still exhibited the childlike curiousity and youthful athleticism that had made him a legend in the underwater world.

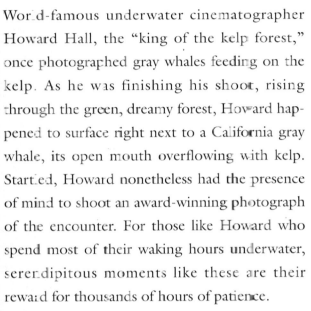

67 bottom The Pacific electric ray *(Torpedo californica)* is capable of producing an electric shock of up to 80 volts. Santa Rosa Island, California. Depth: 10 meters.

68 A gas-filled bladder at the base of each giant kelp blade holds the long, trailing stalks in sunlit water near the surface. Santa Catalina Island, California.
Depth: 20 meters.

68-69 Common bat stars *(Patiria miniata)* crawl among kelp holdfasts and colonies of red algae in a California kelp bed. Point Conception, California.
Depth: 18 meters.

69 bottom This five centimeters long kelp snail *(Norrisia norrisi)* grazes its way up the full length of a kelp plant, removing a thin layer of cells along the way. When it reaches the top, it releases itself to free-fall to the bottom, where it begins its food gathering journey once again. Santa Catalina Island, California.
Depth: 10 meters.

THE GIANTS OF AUSTRALIA
LOOKING FOR THE WHITE SHARK

Underwater photographers are a small society of enthusiasts who readily share their favorite techniques, dive spots, and subjects even as they compete for assignments. Stan Waterman, a gifted underwater filmmaker for 50 years, told me that I must go to Australia to photograph whale sharks and the infamous great white shark. He told me about the guru of great white handlers, Rodney Fox. Rodney knows "The White Death" better than anyone alive. In 1963 he was attacked by a great white during an Australian spear fishing tournament. Since that time his attitude toward the species that nearly killed him has evolved from revenge to reverence. Great Whites move into the waters surrounding Dangerous Reef off the coast of South Australia during the months of February and March. This is when Australian sea lions give birth on the rocky islands. The irresistible scent of afterbirth draws them from many miles away. Even before we anchored, Rodney was ladling out buckets of chum, a foul-smelling mixture of tuna and horsemeat. Within five hours our first dinner guest arrived. It swam around the boat, cautiously inspecting this iron monster that smelled so good. It actually poked its head above the water and studied us with its black, unfathomable eyes. All the photographers on board scrambled into full wet suits and tanks. Grab-

bing my cameras, I leaped off the back end of the boat directly into a sturdy, specially-built shark cage. We drifted 20 meters from the dive boat. The water was a murky, blueish-green. I could just make out a one-kilo tuna head floating nearby. Rodney had hung it there like an underwater doggy-bone. After 15 minutes of silence, I let my guard down, thinking that our subject had decided against a visit. At that moment Mr. Big appeared out of the gloom. Great whites, I would come to appreciate, have the uncanny ability to come up on your blind side. Like a magician, the great white suddenly appears where moments before there was just blue water. This shark swam toward me with effortless grace and confident decisiveness. Ever so slowly, it seemed, it moved closer and closer. As it arched up to grab the tuna head I could see two enormous claspers on its underside that revealed this was a male. His nose brushed the bait. With a snap of its powerful tail, it made a circle around his prey. As he circled, he opened his enormous maw. I can never forget that image. At that moment I realized that, while there might be 370 different species of shark, this shark was in a class by itself. We had enormous luck that trip. We had at least one, and at times as many as five, great white sharks circling the boat for eight straight days. I had plenty of time to observe this impossibly gorgeous animals.

70-71 We worked the whites *(Carcharodon carcharias)* for eight straight days, at times there were as many as five separate individuals circling my cage. Dangerous Reef, South Australia. Depth: 1 meter.

72-73 The white shark is distinctive in look - it is not easily mis-identified. It possesses an enormous black eye that looks right through you.
Dangerous Reef, South Australia.
Depth: 1 meter.

74-75 Once you see Mr. Big you can never forget him. The Great White, is definitely the "Cadillac" of sharks in my book. It is not the length, but the girth, personality and power this fish commands that makes it the king of sharks.
Dangerous Reef, South Australia.
Depth: 1 meter.

THE WHALE SHARK

76-77 The Spring on
Ningaloo Reef in West
Australia attracts
hundreds of whale sharks
(Rhincodon typus) to feed
on the upwellings of
plankton from the deep
nutrient-rich waters
found there. These
biggest of fish awaken
the awe of any human in
the water with it.
Ningaloo Reef,
West Australia.
Depth: 10-25 meters.

Two months after meeting the great white shark, I teamed up with Rodney Fox again, this time to photograph the whale shark migration off Australia's western coast. Every April these giants of the sea return to Ningaloo Reef for the free meal prepared by the upwellings and plankton blooms that spur the growth of shrimp-like krill and similar fare. As with the great whites, Rodney had perfected an ingenious way of locating these elusive giants. Since whale sharks frequently cruise close to the surface, it was an easy matter for a spotter plane to direct us via radio to within 100 meters of the animal. The dive boat would then drop us off directly in the projected path the animal would take. Our luck was not as good as it had been on the great white dives. During the first four days, bad weather prevented us from even getting close to the animals. The one time our spotter did locate a whale shark, it dove before we ever got a chance to see it. But the fifth day dawned with a clear sky and a calm sea. Within minutes after take-off, the plane spotted a medium-sized whale shark cruising slowly just beneath the surface. We strapped on miniature dive tanks to reduce weight and drag so we could swim fast enough to keep up with these beasts. Cradling two of my widest wide-angle lenses, I braced for the signal... It came.... Go... Go... Go! The shark was upon me before I had even cleared the water from my

mask, moving at a leisurely pace, with its mouth partly open. Years of photography in the sea have prepared me for moments like this — instinct takes over. My right eye was clamped shut; my left was glued to the viewfinder. I held my ground until I thought a collision was inevitable. With less than 5 meters separating us, I let the shutter pop. I took my eye from the camera and focused on this blue-spotted giant silently approaching. I braced myself. When she came to within three meters she eased her enormous bulk gently over and above me. As its dorsal fin went by, I reacted instinctively. Grabbing hold, I gently hung on. With my left hand clinging to its fin, my right was busy snapping photos of the enormous mass in front of me. The shark began to dive. It chose a low angle of descent and didn't seem to mind my presence. At a depth of 40 meters, I let go. As I slowly finned toward the distant surface I watched this gentle female melt into the distant blue. When I surfaced, I was two kilometers from the dive boat. As I bobbed in the water waiting to be picked up, I could not remember a more satisfying encounter with a marine animal. For the next five days, Neptune was in my corner. Not one day went by without at least five whale shark dives. On the final day, I jumped into the water and chased whale sharks no less than twenty-two times. On the way back home that night, I slept.

78-79 This photo has stopping power! Many tons of stopping power. Taken within meters of its presence, the mouth of this whale shark impresses all but the plankton it consumes.
Ningaloo Reef,
West Australia.
Depth: 5 meters.

80-81 Sometimes, if done gently and with the right individual, you can approach a whale shark as close as you dare without arousing fear on the part of the fish.
Ningaloo Reef,
West Australia.
Depth: 10 meters.

GALAPAGOS

82-83 This marine iguana *(Amblyrhynchus cristatus)* is the only lizard in the world that can be found in the ocean. Waiting on the rocks for low tide, it swims into the shallows in search of its algae meal that it rips from the intertidal rocks. Fernandina Island, Galapagos.
Depth: 3 meters.

In 1985, the German magazine *GEO* funded a project for me to shoot an underwater feature on the secret world under the temperate waters of the Galapagos. Straddling the equator a thousand kilometers due west of Equador, more than 50 islands, some no more than rocky outcrops, make up the Galapagos Islands. While much of the land is desert, the sea is as rich an environment as any rain forest. The Galapagos Islands are nourished by a rich recipe of converging currents. Frigid currents from the Antarctic and tropical currents from the equatorial Pacific create a congenial meeting place for cold and warm-water marine species. Diving in the Galapagos has all the elements of entering The Twilight Zone, and you quickly learn to expect the unexpected. What may start out as a routine dive quickly turns bizarre, as you pass a parade of fantastic and diverse marine animals that would normally only appear together in a world-class aquarium. Big animals make their appearance on every dive. For instance, you might be examining sea anemones along a rock wall, when you turn to see a school of thirty hammerheads cruising toward you. Or perhaps some instinct tells you to look up on an early morning dive. You are stunned to see a squadron of bat rays flying in perfect formation against the sky-blue surface. Sea lions perform submarine acrobatics all around you, as if to emphasize your physical limitations. Giant sea turtles practically bump into you. They simply do not perceive you as a predator in these protected waters. For me, the oddest sight was the diving marine iguanas. Nowhere else in the world are these marine reptiles found, except in the evolutionary isolation of the Galapagos Islands. Each day, at the single low tide of the day, the iguanas abandon their sunny perch on lava ledges and venture into the crashing surf to rip a meal of algae from the intertidal rocks. In order to get within picture-taking distance, I had to strap on an extra-heavy weight belt to keep from being swept away by the swells. I remember trying to photograph one hungry individual feeding in the surf as I was being tossed back and forth by the waves. I felt like I was in the spin cycle of a very powerful washing machine.

84-85 The sea lions
(Zalophus californianus
wollebaeky) in the
Galapagos are everywhere.
Playing like puppy dogs,
they seem to mock a
divers awkwardness.
Daphne Island, Galapagos.
Depth: 4 meters.

86-87 A school of bat
rays (Myliobatis sp.) in
flight formation was my
reward for entering the
sea at sunrise. Photo
opportunities are
multiplied at sunrise and
sunset-these changeover
periods explode with
action.
Targus Cove, Galapagos.
Depth: 20 meters.

88-89 The red-lipped batfish *(Ogcocephalus darwini)* is commonly found in the Galapagos waters. The best time of appoach is at night under the cover of darkness. Targus Cove, Galapagos. Depth: 15 meters.

89 top A juvenile scorpion fish is caught in sleep on the dark volcanic sand of a protected cove. Targus Cove, Galapagos. Depth: 7 meters.

The ocean floor also holds many surprises. One of the strangest fishes I have ever seen makes its home in the volcanic sand bottom of the Galapagos: the red-lipped batfish, it looks as if a Hollywood make-up artist got carried away. Thick, ruby-red lips and a hornlike projection on its forehead give it an arresting appearance. Its pectoral fins have evolved into leg-like appendages that allow it to balance on its fins and tail like a tripod. Despite being on the equator, the seas surrounding the Galapagos Islands are decidedly temperate, even downright cold. The combined currents support a wealth of suspended plankton that nourishes underwater life and frustrates underwater photographers. More than 10 meters of visibility is a treat, but I am happy to accept these limitations in order to savor "the unexpected" and the opportunity for a magical surprise to suddenly materialize out of the murky depths. One day, schools of amberjacks might swirl around you in dizzying circles until your air runs out; another time, they pass you by without a backward glance. Or you might be mauled by a "classroom" of sea lions, not merely 5 or 10, but a crowd of 30 or more eager attention seekers. The Galapagos is justly famous for its terrestrial phenomena. Just imagine the same underwater, but with no one else looking over your shoulder while you experience it. Don't miss it!

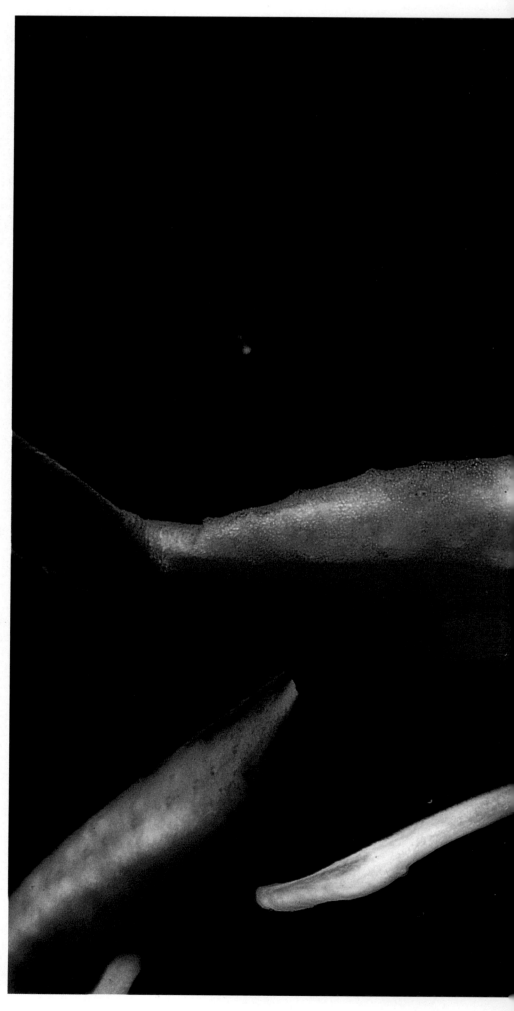

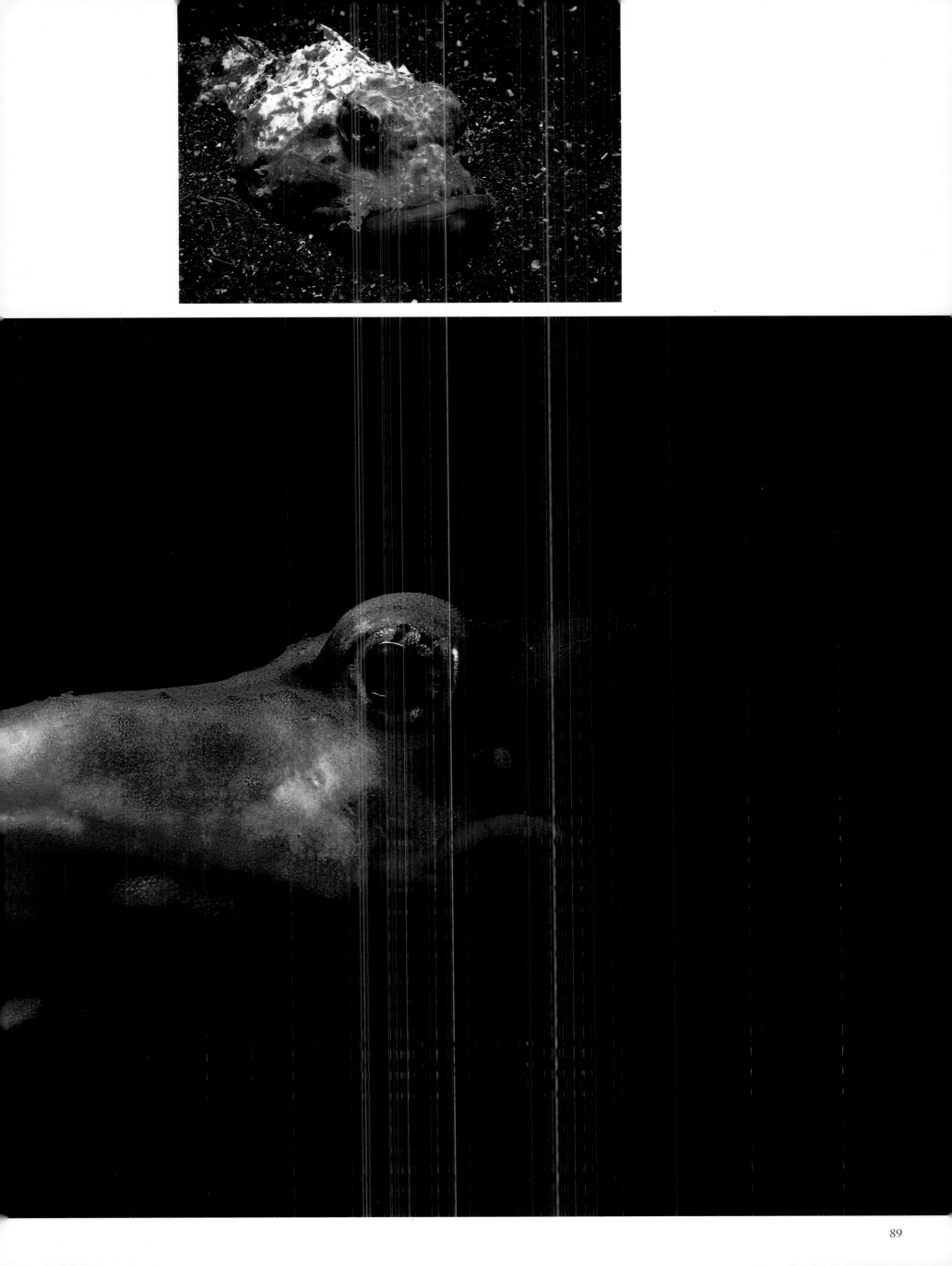

90 Predators themselves, bristleworms arm themselves with fragile, needle-sharp spines that break off in a larger predators' flesh. Galapagos Islands. Depth: 15 meters at night.

91 The Guineafowl puffer (*Arothron meleagris*) is usually with a black body and white spots. When you get lucky you find the golden yellow phase. Kicker Rock, Galapagos. Depth: 5 meters.

92-93 Jointed mouthparts at the ready, and antennae aquiver, this hermit crab searches for food on the night-time reef of the Galapagos.
Española Island, Galapagos.
Depth: 5 meters.

94-95 Galapagos diving is best described as "expect surprises". One minute you can be looking out into blue water, the next minute you look up and are surrounded on all sides, bottom, and top by a school of jacks *(Seriola sp.)*.
Galapagos Islands.
Depth: 15 meters.

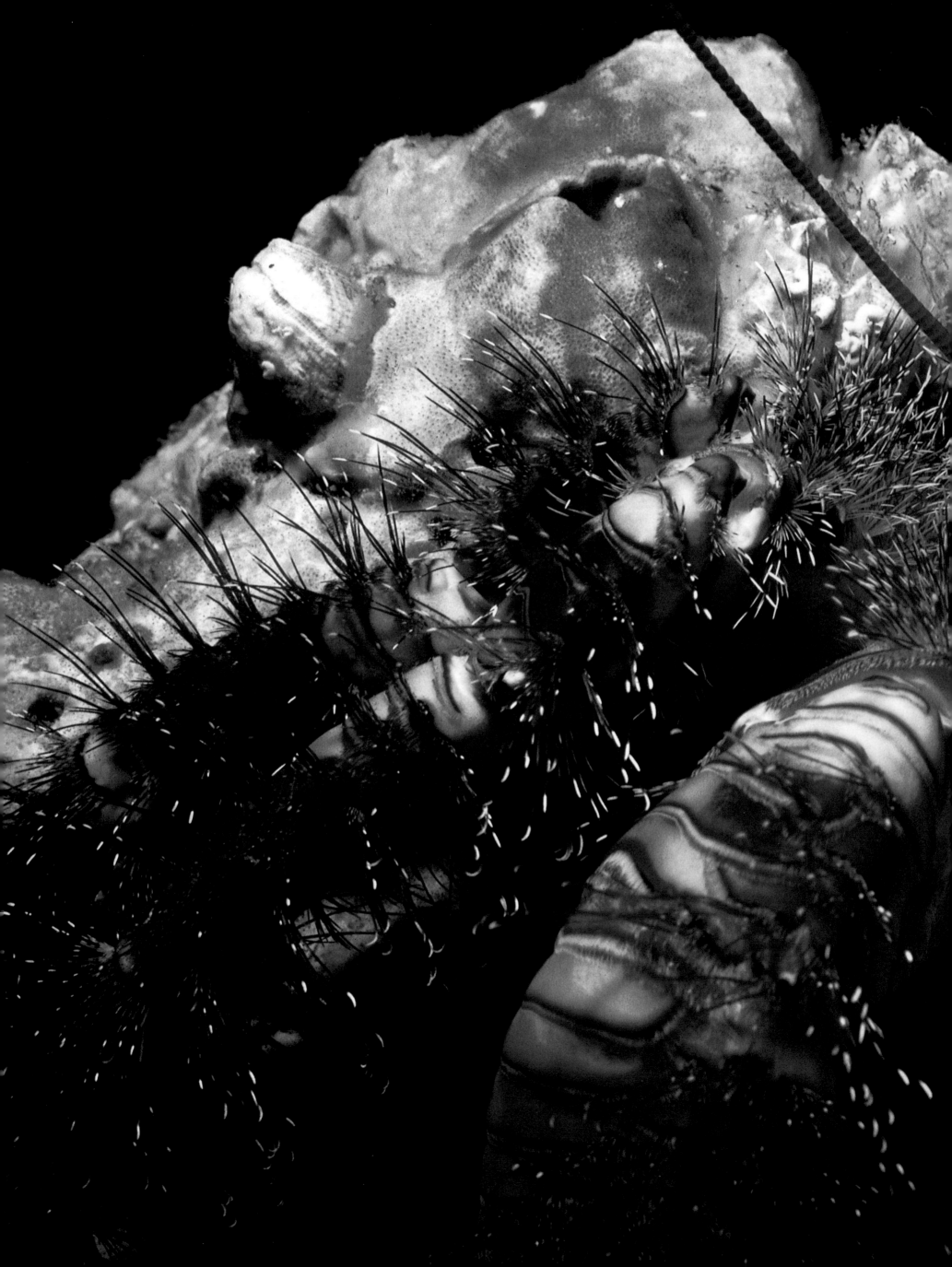

PAPUA, NEW GUINEA

In February, 1995, I was searching for a spectacular and unique coral reef that was off the beaten track. I found it in the volcanic arc of Kimbe Bay in the Papua New Guinea province of West New Britain, in a dive paradise known as the Walindi Plantation. Max and Cecile Benjamin have created and run a dive resort next to some of the finest diving anywhere in the world. Its only drawback, but probably also its saving grace, is that the travel time to get there from "civilization" is measured in days, not hours. My companion was dive legend Oren Lifshitz. Because of the knowledge he has gained reconstructing reefs, few people understand the balance of a coral reef as well as this man does. After our first day on the Walindi Plantation reef, he summed up the whole experience in two words — wild and magical! What first struck us about this reef was how incredibly healthy and pristine it was. Although it had one of the most diverse "collections" of reef fish I had ever seen, no one species appeared to have to compete for living space. An incredibly fast-growing reef, nurtured by river runoff and remarkably warm water temperatures, constantly added new turf for the reef residents. A rich planktonic soup reduced visibility, but grounded the complex food web of the reef. The lack of commercial fishing and land-based pollution also nurtured the reef.

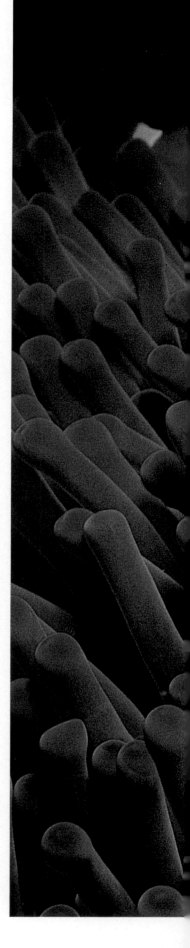

96-97 The most famous case of living together on the coral reef is the easily visable and endearing partnership between the spunky clownfish and the sea anemone.
Walindi Bay,
Papua New Guinea.
Depth: 4-12 meters.

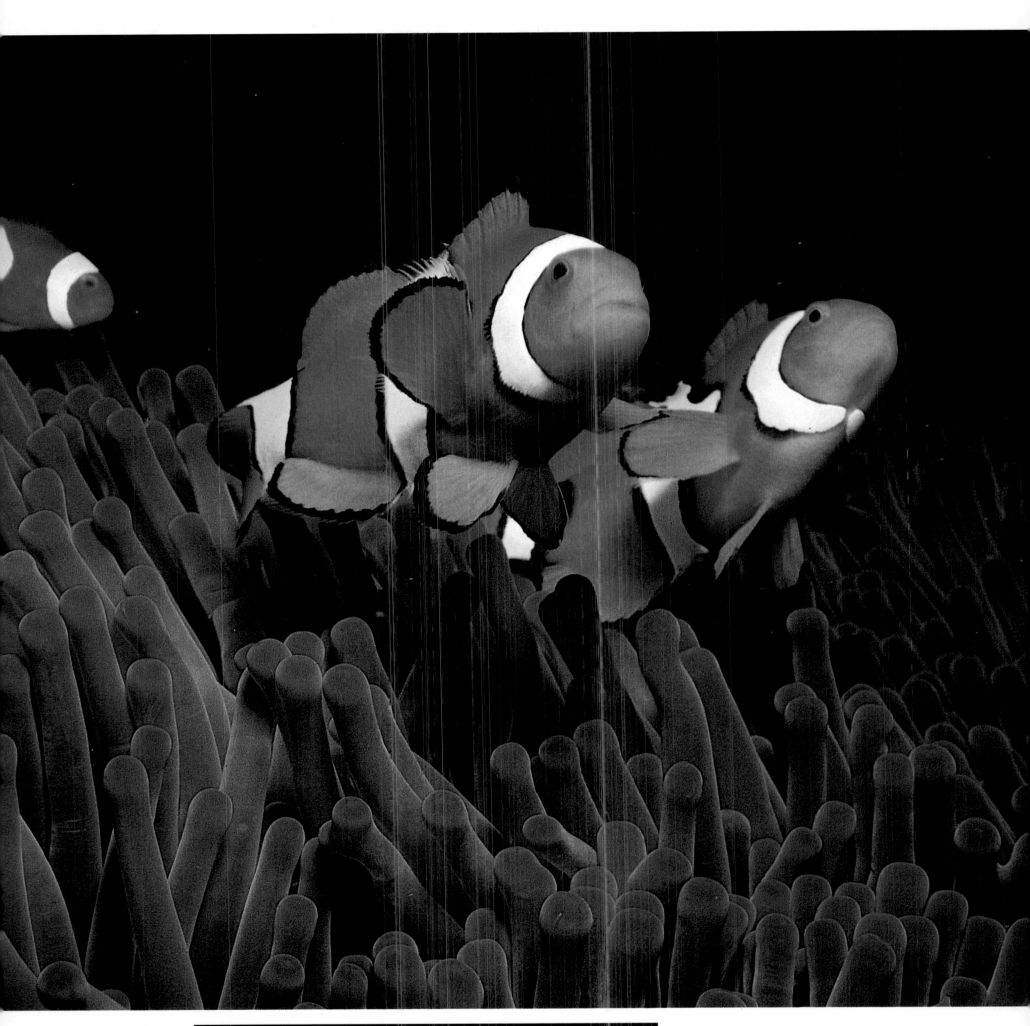

98-99 Tightly contracted, these anemones have expelled much of the water that keeps them erect. In this state they use little metabolic energy and rest in relative safety.
Walindi Bay,
Papua New Guinea.
Depth: 6 meters.

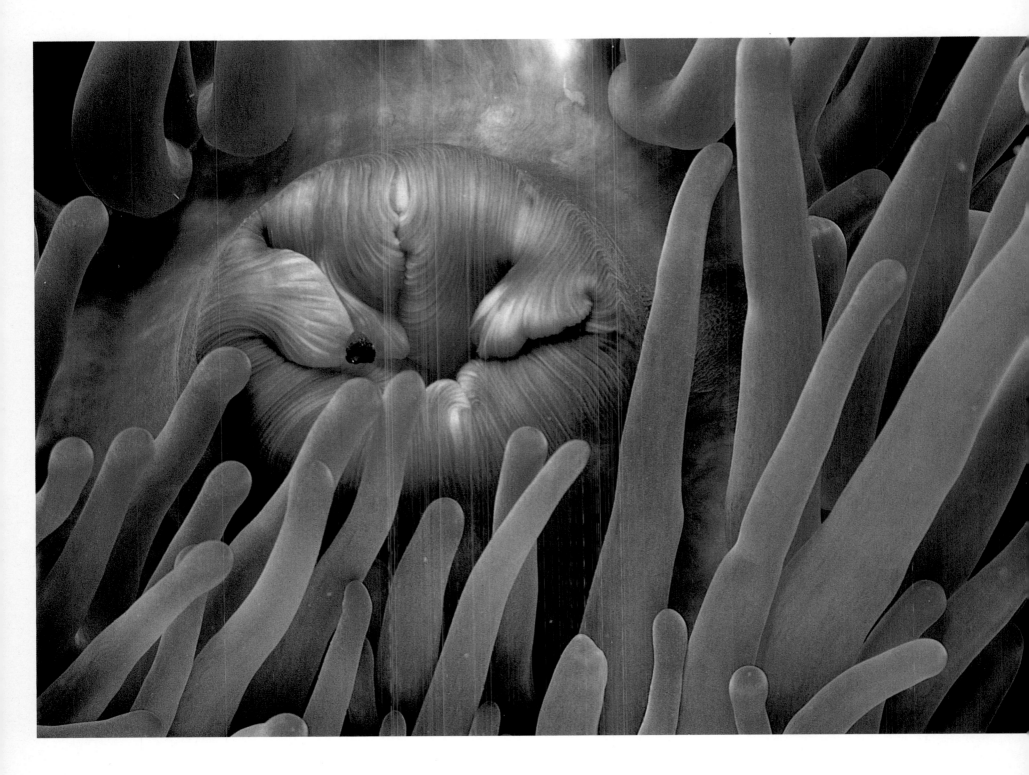

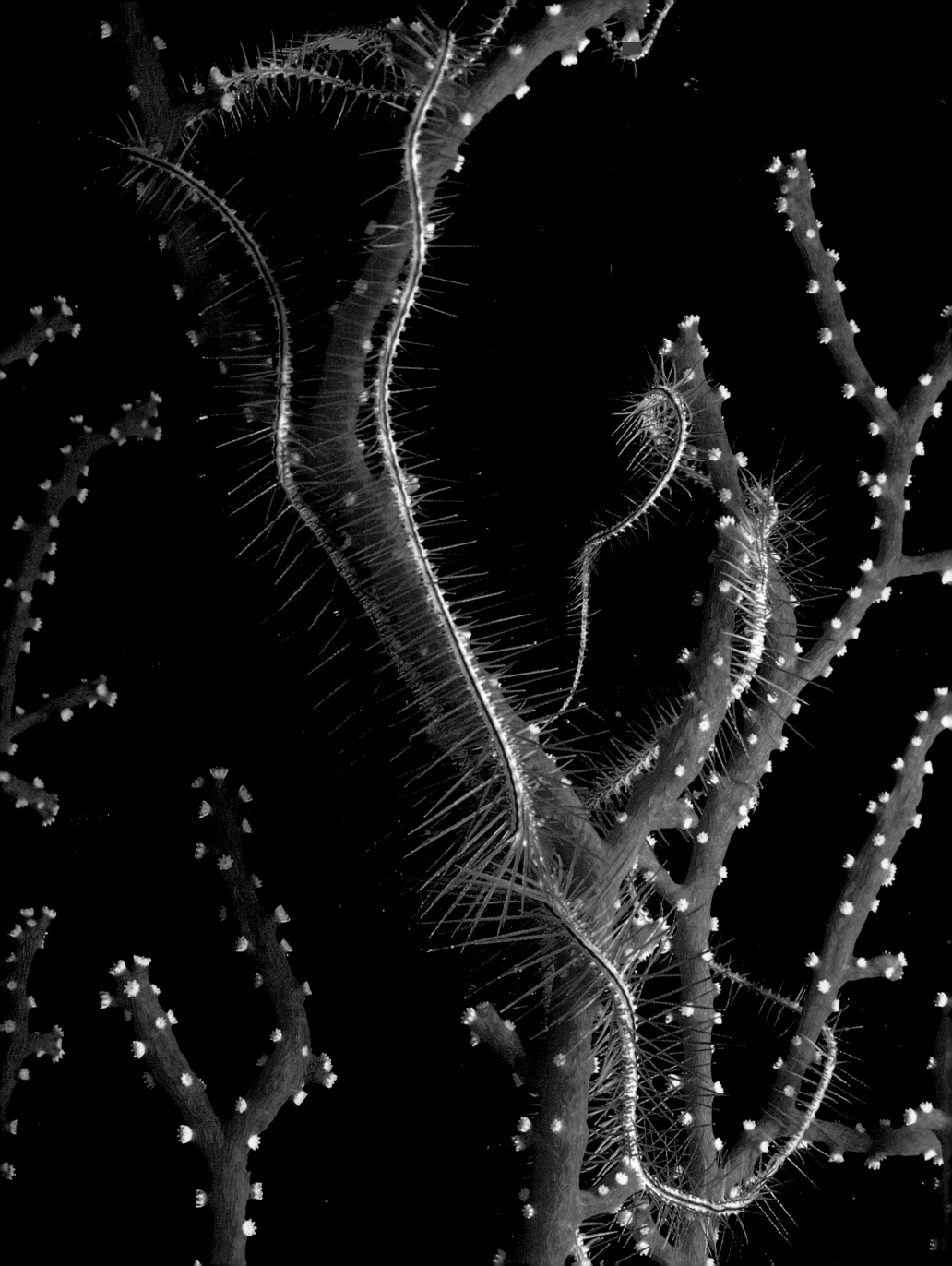

The marine topography of this region is a combination of fringing mainland and island reefs, offshore reefs, and deep-water pinnacles.

These isolated reefs in the deep ocean appear to have formed just as Charles Darwin theorized in the 1860's. Corals grew upward from the slopes of extinct volcanoes as the lava cones slowly sank into the sea.

Oren, a man not given to excessive compliments, felt the stony coral formations here were unparalleled, even in comparison to the world-famous reefs of his native Red Sea.

A recent scientific survey of the marine fauna around Walindi Reef confirmed the rich biodiversity that we observed on our dives.

The survey identified over 320 species of hard corals, representing over half of the world's coral species in this one small area. Over 700 species of fish have been identified to date, including more than twenty species of butterflyfishes at Kimbe Island alone. In the space of twenty meters, it is possible to find fishes ranging in size and diversity from delicate lagoon damselfishes to larger unicornfish to ocean-going sharks.

Each reef has a distinct personality that makes it memorable and special, shaped by the plants, animals, and environmental conditions unique to that locale. So it was with Walindi.

We were there during the rainy season, so my memories of this reef will forever include the five downpours each day, accompanied by the fireworks of lightning and thunder, followed by a misty calm that would settle over Kimbe Bay and beckon us to enter its waters. Responding to the invitation, we would don our thin skin suits. I would be greeted by pulsating life every time I entered the water. Giant plate corals fanned out like open umbrellas turned toward the sun, providing abundant hiding places for reef creatures. Coral fans and gorgonian sea whips bent with the current, their tiny polyps grabbing plankton from the water twenty-four hours a day.

One day, Oren led me to a school whose membership included thousands of barracuda. This wild community accepted me, or at least tolerated me. Over the course of the day, I merged with the school, almost joining with the mass of eyes, scales, and slender shapes. This massive cloud of silver arrows shot back and forth, up and down through the water, dazzling me with their silvery light. I spent myself with them as one might spend oneself in love. Then there was the curious schooling behavior of catfish balls, tiny, poisonous fish, each one no more than a thin sliver three centimeters long. Together they became a tightly-packed ball of thousands of aggressive individuals, buzzing me like angry bees.

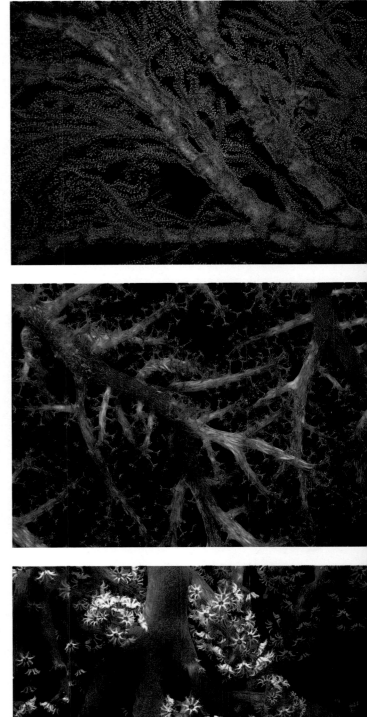

102 This crab was spotted at night on an alcyonarian soft coral. With a touch of backlighting its structure outlines the coral. Kimbe Bay, Papua New Guinea. Depth: 7 meters.

103 Standing slightly bent by the current, gorgonia wire corals add a graphic element of vertical lines to the reefs of Papua. Kimbe Bay, Papua New Guinea. Depth: 12 meters.

104-105 This gorgonia coral falls in between hard and soft coral in terms of its body structure. It lacks the limestone skeleton of hard coral, it does produce an inner core of tough, flexible protein akin to human hair or fingernails. Kimbe Bay, Papua New Guinea. Depth: 12 meters.

106-107 and 107 bottom
On spotting a potential
meal, lionfishes *(Pterois sp.)*
use fanlike pectoral fins to
corral their quarry in a
tight corner. A lightening
fast gulp of a large,
vacum-cleaner-like mouth
then often spells the end
of the hunt.
Kimbe Bay,
Papua New Guinea.
Depth 7 meters.

106 bottom Triggerfish,
strong swimmers with
powerful jaws, feed on
small crustaceans and
echinoderms both on and
around the reef. In Papua
every reef I visited had at
least one pair of this type
of triggerfish *(Balistoides
viridiscens)*.
Kimbe Bay,
Papua New Guinea.
Depth: 8 meters.

The zaniest-colored fish I have ever seen lives on Walindi Reef. The clown triggerfish, with bright yellow lips and garish spots and patterns everywhere, well deserves its name.

This sometimes hefty clown was quite approachable. It didn't seem to mind being followed as it meandered across the coral, pausing every few meters to take a bite out of the reef to satisfy its considerable appetite.

There were other outrageous displays of color on the reef. Perched atop barrel sponges large enough to hide a diver, feathery crinoids presented a riot of color and design.

Clouds of deep-purple anthias were buffeted by the currents that passed over the reef. Oren believed their unusually large concentrations were due to the abundance of living spaces available on the reef. Even getting to the reef was a delight of distractions.

On our way out to the reef one day our boat was surrounded by a pod of pilot whales that seemed to demand our attention. We obliged by snorkeling with them for half an hour before resuming our journey.

On other occasions, people have been diverted by dugongs, orcas, whale sharks, sailfish, and even once by a sperm whale.

Getting back to shore was not always so much fun. Our captains, Joe and Dominique, who grew up in the village down the road from Walindi Plantation Resort, sometimes shared a "good chew" of middle intoxicating beetle nut.

One afternoon on our way back to shore, heavy rains darkened the sky. Joe asked me to shine a light on his compass, so he could orient the boat straight toward home port only a kilometer away. He was relaxing on the bow when a coral reef mysteriously leaped in front of the boat, sending a fiberglass platform with my eight cameras into space. Oren was hit by our lunch leftovers of overripe papaya slices. Captain Joe looked skyward for help, while Oren looked down at the boat trying to determine the extent of the damage.

Captain Joe assured him that everything was okay. "Hey, this is the ocean. Things go wrong out here." Because of our attack by the marauding reef, we were a little late getting in that evening.

Walindi Reef is a must-see marine environment for coral reef divers. It looks like it may remain the utopian place we found it.

Scientists have found little potential here for a commercial fishery for fish or for invertebrates such as trochs, clams, and sea cucumbers.

The economic potential of the area lies in protecting the great diversity of the coral reef ecosystem to promote ecotourism.

Good diving!

108-109 Levers of bone and hinges of cartilage extend this grouper's mouth *(Plectropomus oligacanthus)* into a tube, enabling it to engulf and trap its bite-sized prey within a fraction of a second.
Kimbe Bay,
Papua New Guinea.
Depth: 7 meters.

110-111 This five-lined cardinalfish *(Cheilodipterus quinquelineatus)* was surprised sometime after midnight by my intrusion. In the background of soft coral a brittle star can be seen making its way through the night.
Kimbe Bay,
Papua New Guinea.
Depth: 9 meters.

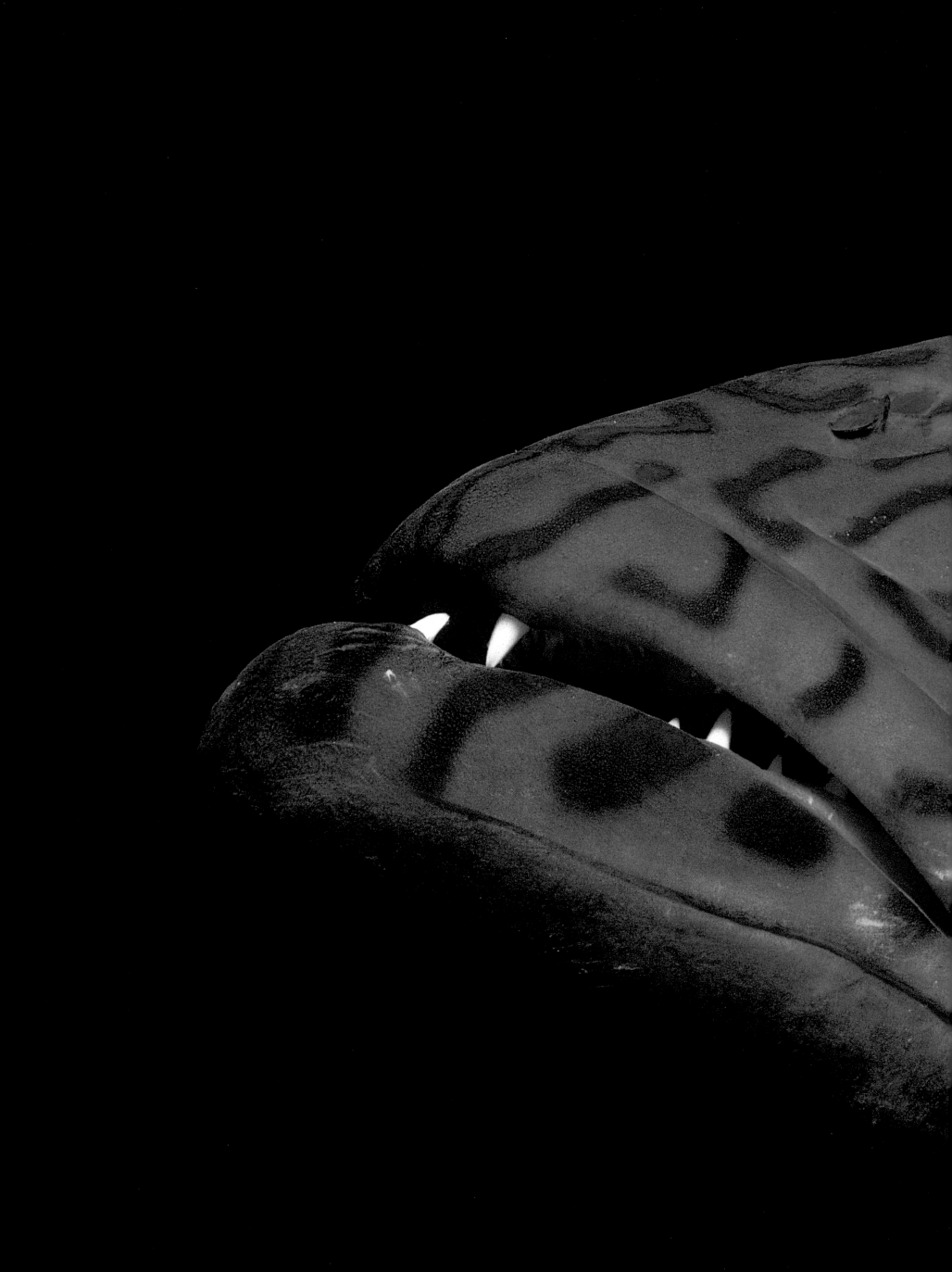

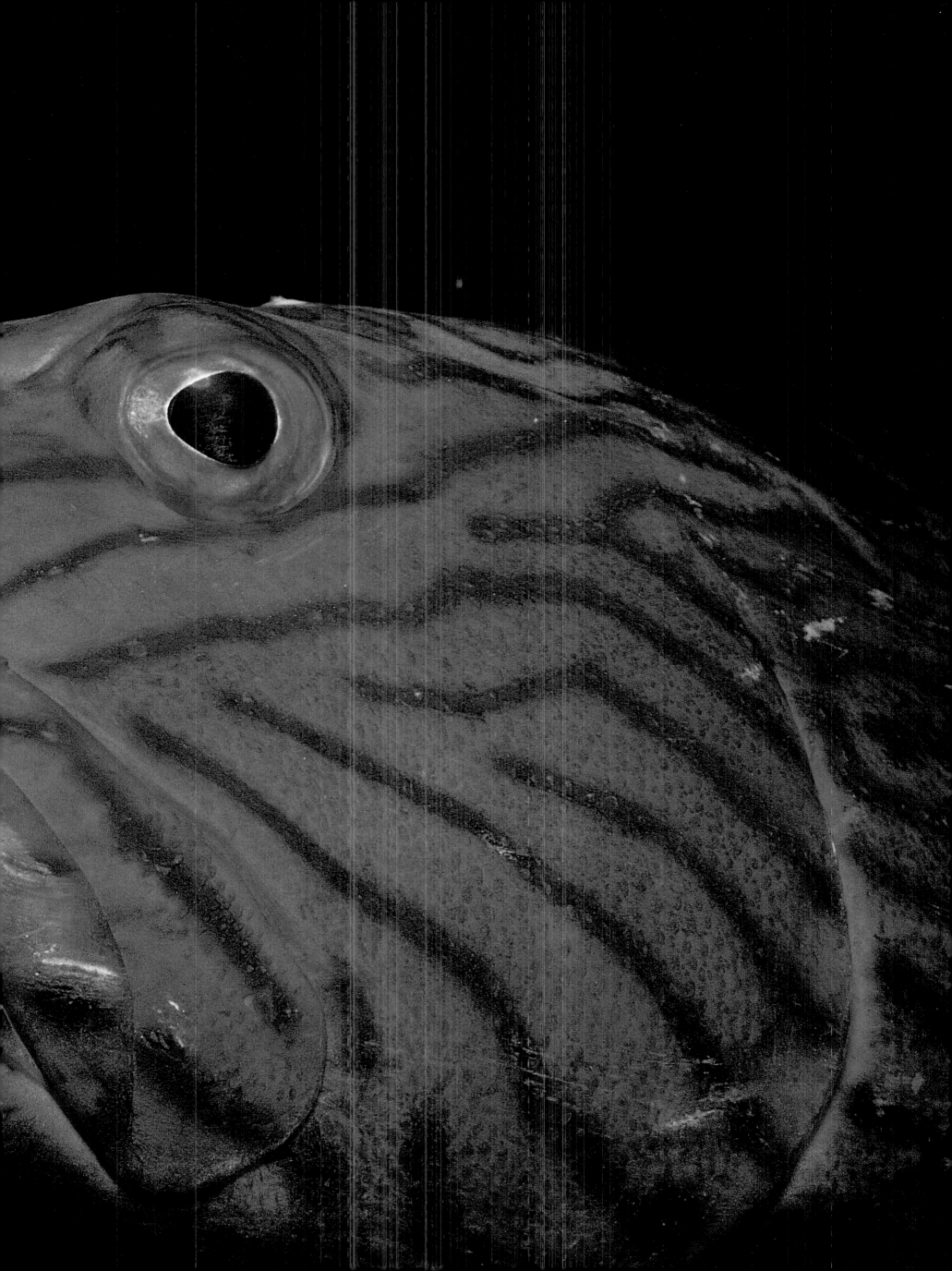

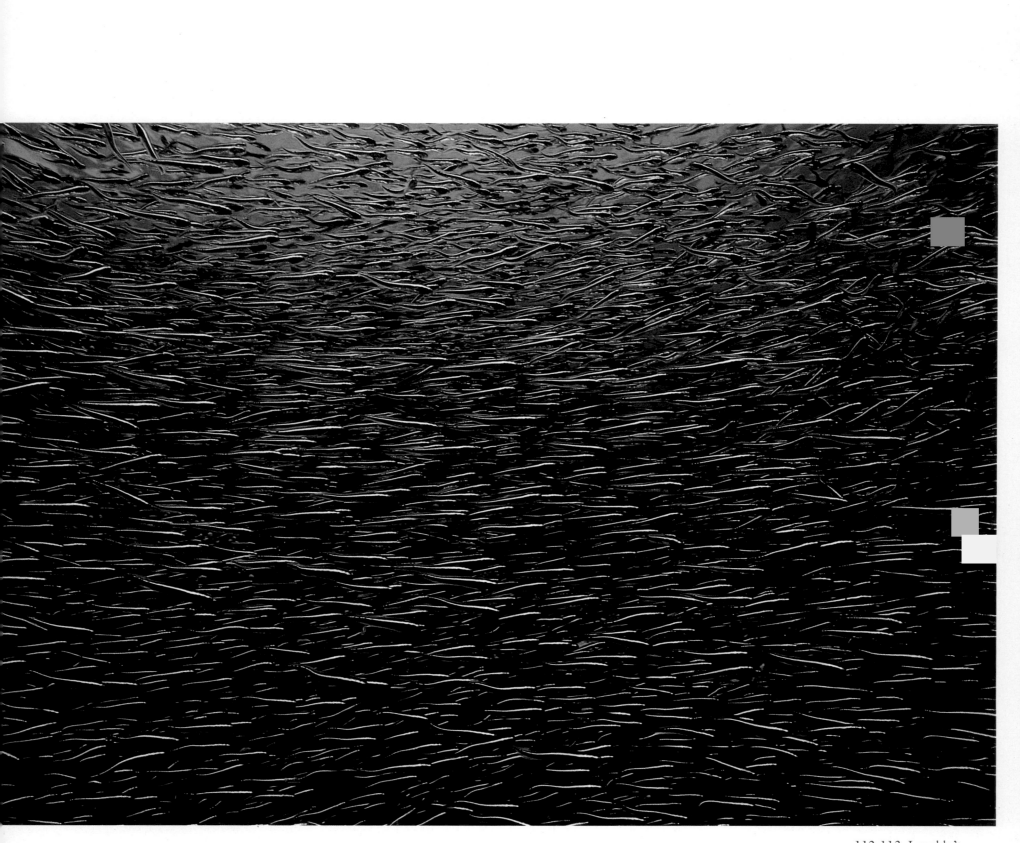

112-113 I couldn't believe my eyes, a giant school of venomous catfish *(Plotosus lineatus)* was swimming in their ball-like fashion. The school would split and rejoin many times over the course of my dive. Kimbe Bay, Papua New Guinea. Depth: 16 meters.

114-115 I got lucky in gaining my acceptance to a resident school of barracuda *(Sphyraena sp.)*. First came apprehension - what form of strange predator was I? Perhaps just an overgrown, retarded remora. In the end I don't know quite how this school thought of me. I spent 12 hours during a two-day period exclusively with them.
Kimbe Bay, Papua New Guinea.
Depth: 15 meters.

116-117 I was told there was a drifting coconut palm headed out to open sea. Floating in an open sea, almost anything becomes an island of refuge. Here a school of juvenile golden trevally jacks *(Gnathanodon speciosus)* find food and shelter.
Kimbe Bay, Papua New Guinea.
Depth: surface.

118-119 Purple anthias *(Pseudanthias tuka)* are characteristic fish of the reefs of Papua New Guinea. Their purplish coloration adds a spark to its reef.
Kimbe Bay, Papua New Guinea.
Depth: 18 meters.

SIPADAN

In 1991, as I soaked up the old world splendor of Vercelli, town situated in the middle of the Po Valley, in Italy, I listened to Carlo De Fabianis and Marcello Bertinetti sketch out a proposal to explore a place that I had never heard of: Sipadan Island. On a recent trip to Singapore, they had overheard local divers describing it as a lost "Garden of Eden." Allegedly, it had the most amazing population of turtles that had somehow escaped the plunder of turtle poachers and egg thieves. Would I join them diving in Sipadan?

Within short order, in Borneo, we filled a speedboat full of enough professional still and video equipment to justify as pirate booty. For two hours, we raced across the mirrored surface of the Celebes Sea until we reached an underwater seamount. On its Pacific side, the volcanic island plunges thousands of meters to the ocean floor. Glistening white coral sand, lush vegetation, and minimal evidence of human civilization made this tiny tropical outpost a place where Robinson Crusoe would have felt at home. Within minutes, we were ready to dive. Marcello frantically signaled me to join him. Perched at the entrance of a sheltered crevasse, he was watching two giant green turtles mating with abandon. This is the type of photographic opportunity that *paparazzi* and underwater photographers dream about. I shot roll after roll of the romantic couple. When I left, the lovers still hadn't parted or surfaced for air. Back on the boat, the first words out of Marcello's mouth were, "I get 10% royalty on the sales of that photo!" He was right; he had handed me a great shot.

That evening we explored the smooth, perfect beach surrounding the island of Sipadan. There were turtle nests everywhere. We followed some tracks in the sand until we found a female turtle at the high tide line, straining to release her eggs. For more than an hour she slipped a hundred ping-pong ball sized eggs into a depression in the sand. Then she covered the eggs with sand, patted the surface firmly with her shell, and scattered more sand around with her flippers to disguise the nest.

The following morning we entered the water early and were greeted by a wall of jacks. Within three minutes, we spotted a turtle. The turtle seemed unfazed by Carlo's video camera. Only when he turned on his powerful movie lamps did the turtle dart away. Though turtles are clumsy on land, they are surprisingly quick and agile underwater. Meanwhile, Marcello was out in open water swimming next to a gigantic male that held its characteristically long tail straight out behind it. We were amazed that it allowed Marcello to come so close. These turtles didn't harbor the fear of humans that sea turtles elsewhere exhibited.

120-121 After reaching the shore, a young turtle swims in the dark open sea, ignoring the dangers and the predators that might meet.
Sipadan, Borneo.
Depth: 10 meters, at dusk.

121 top Thanks to the torch light I discovered in the reef a couple of turtles *(Chelonia mydas) while mating.*
Sipadan, Borneo.
Depth: 10 meters, at dusk.

122-123 Light will disappear within a half hour. This turtle searches for its evening hotel room on the reef in which to spend the night. Sipadan, Borneo. Depth: 12 meters, sunset.

124-125 As the sun drops and the changeover begins packs of jacks on the hunt are cruising the drop off ledge. Sipadan, Borneo. Depth: 8 meters.

By the end of a day following sea turtles, Marcello established the record for finding the most, thirty-eight on one dive! That night I waited until it had been dark for three hours and entered the water alone. On a ledge four meters below the surface, I spotted two giant 20-kilo bump-headed Parrotfish fast asleep. During the day that same niche had been a grazing area for the seemingly limitless turtle population. I descended down a vertical wall of coral that was full of life and nocturnal action.

I directed my lamp toward what appeared to be two eyes poking out from the sand and discovered a perfectly-camouflaged juvenile crocodile fish no larger than my thumb, partially buried in the sand. I chose my angle carefully and started shooting. On one evening we watched more than 100 baby hatchlings emerge from their shells and scramble down the sandy slopes to the sea. On another dive we discovered a turtle graveyard — a cave of turtle skeletons. Most probably, the air-breathing reptiles had not been able to find their way out of the tunnels.

We felt very fortunate: we were among the first foreign divers to have seen the unspoiled splendor of the island of mating turtles. Yet I feared that Sipadan's discovery would be its undoing. There are already far too few Sipadans left.

If you go there, take only photos, and leave only footprints.

126-127 Sipadan, like most dive jewels, has special gifts to offer. One is color - nowhere can you find the strong, bright, electric coloration that cloak sponges and soft corals of Sipadan Island display.
Sipadan, Borneo.
Depth: 4-30 meters.

128-129 While both
these animals are looking
like anemones, only the
reddish one is a true
anemone. The green
with white-frosted tipped
is a coral - the Anemone
Mushroom Coral
(Fungia sp.).
Sipadan, Borneo.
Depth: 25 meters.

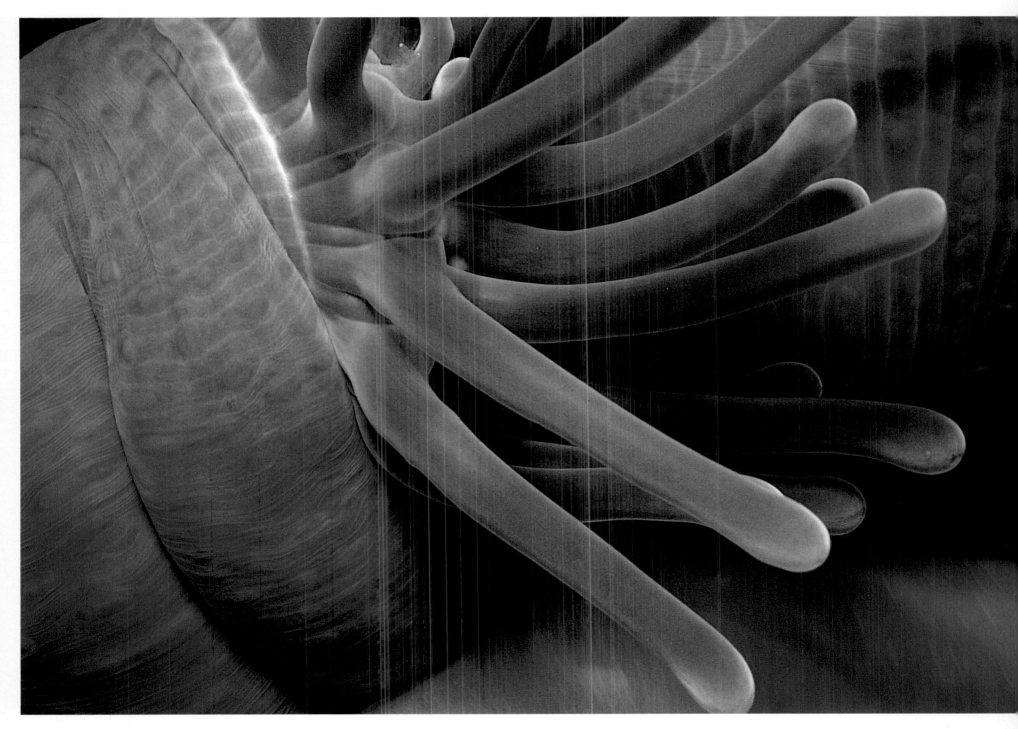

130-131 With a bright yellow coloration it is not difficult imagining why one might name these "Clownfish" (Amphiprion percula). Sipadan, Borneo. Depth: 8 meters.

132-133 Carrying venomous spines concealed beneath all that frilly finnage, this lionfish (Pterois antennata) stalks its prey slowly and imperturbably during dawn and dusk twilight periods. Sipadan, Borneo. Depth: 10 meters at dusk.

RED SEA

134-135 Soft corals are
one of the strongest
pleasures in Red Sea
diving - simply put -
the best you will ever see.
Ras Mohammad,
Red Sea.
Depth: 10 meters at night.

I have no closer relation-
ship to any sea than that which I share with
the Red Sea. I have photographed it more
than any other place, and yet it still yields new
discoveries on every dive.

When I first came to dive the Red Sea in ear-
nest during the summer of 1976, I arrived
with only a tank on my back and ten
kilogrammes of lead around my waist.

When I moved over more permanently in
1980, I had scraped together enough money
to buy a beaten up four-wheel drive Jeep and
rent a small apartment in the city of Ofira.
From this base of operations, I was less than
five minutes from one of the world's greatest
thriving coral reefs. Within a year I was joint
owner of a small Zodiac inflatable dive boat.
We named it *Freedom*, because it made us
completely mobile and indipendent. With my
Jeep and my Zodiac, I could go anywhere
along the Sinai coast. Loading twelve diving
tanks and a mountain of camera and camping
gear onto an open trailer behind my Jeep, I
would head out to promising dive sites that
were still completely untouched.

At first glance, many reefs look the same, but
each site has its own special character, partly
defined by its underwater topography.

With practice, as soon as I got the "lay of the
land," I could predict what kind of marine life
I would find at each dive site. Caves and
carved-out terraces located close to areas of
upwellings, are what I like to call "real estate
potential." They frequently harbor glassy
sweepers that tend to school around certain
types of coral and cave entrances. In the open,
more pelagic Straits of Tiran, the reefs of Jackson, Gordon, Woodhouse, and Thomas, big
pelagic fish, sharks, and even dugongs might
make a sudden appearance.

World-famous Ras Mohammad, lying at the
southern-most tip of the Sinai Peninsula, has
some of the most startling seasonal visitors
you can imagine. Here, off an island in Ras
Mohammad during April and May, fish con-
gregate in huge numbers, creating a solid wall
twenty meters deep and ten meters thick.

Possibly because my passion is sharks, some of
my fondest memories of the Red Sea are in the
early morning and late afternoon changeovers
when sharks visit the reefs.

My first encounter with a shark underwater
was on a late afternoon dive off the wall of Ras
Mohammad in 1976. At 15 meters, I looked
down to see a sculpture of strength with an
unmistakable rectangular head. It was a huge
hammerhead. I stopped dead in the water,
stunned, and then impulsively I gave chase.
The shark was aware of my pursuit but totally
unconcerned.

136-137 This coral
grouper (*Cephalopholis
miniata*) displays the
adaptations that make it
such a fearsome predator
during the twilight hours
of dawn and dusk.

Its mouth is large and
equipped with teeth that
enable it to grasp prey
firmly and swallow them
whole.
Ras Abu Galoom, Red Sea.
Depht: 10 meters.

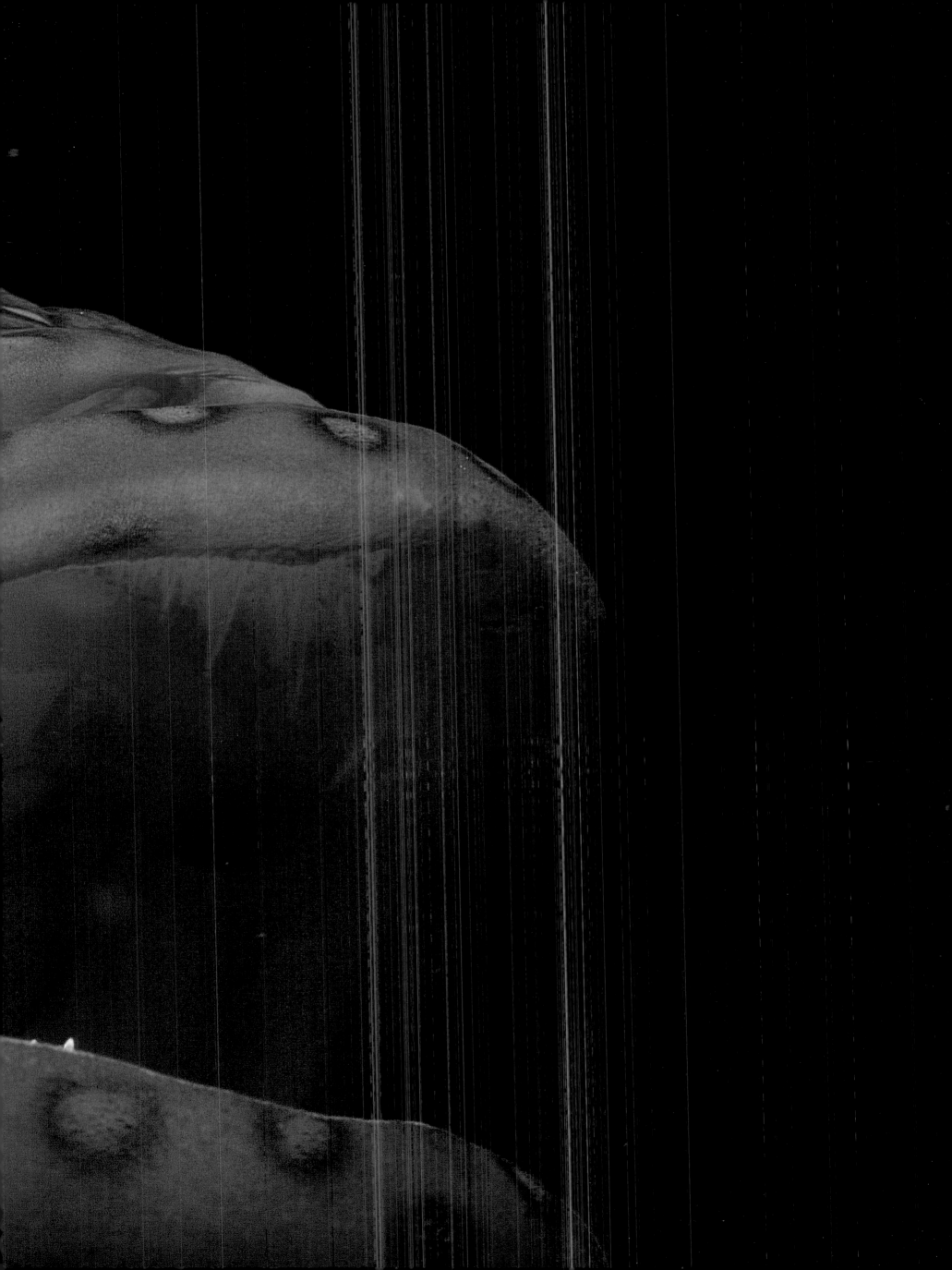

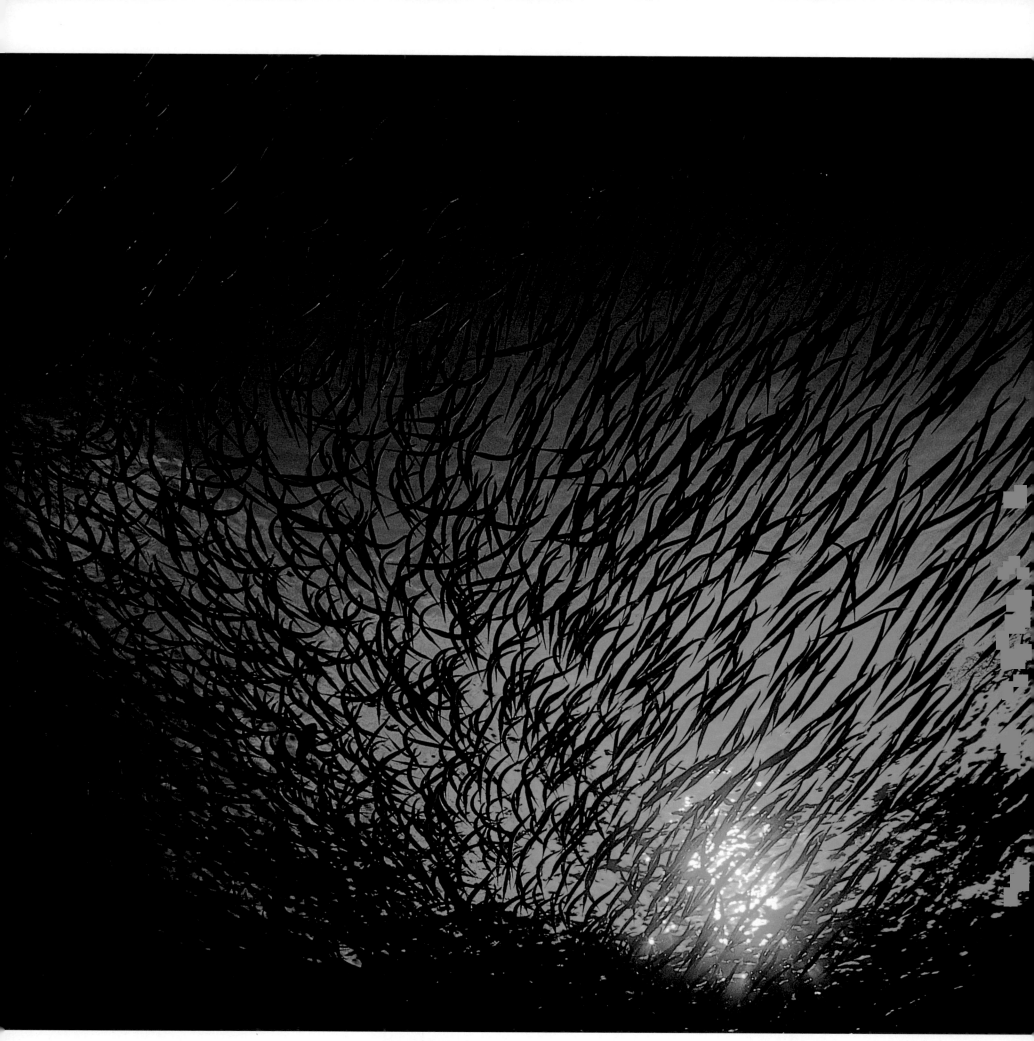

138-139 As day dawned
a school of juvenile
needlefish schooled past:
they were hunting and
being hunted.
Jackson Reef, Red Sea.
Depth: 1 meter.

138 bottom A school of
Red Sea fusilier *(Caesio
suevica)*, swimming close
to the surface, reflects
light into the blue depths.
Gordon Reef, Red Sea.
Depth: 2 meters.

It would let me get only so close before a bolt of effortless strength would send it flying forward at an impossible speed.

It was there I learned my first shark lesson: Don't try to chase the master; the shark will decide if it wants to get close to you.

One morning I awoke before sunrise after sleeping in the lee of Jackson reef. As my dive buddy and I sipped sweet turkish coffee we were aware that the rising sun would soon cause a commotion on the reef below us.

We wanted to be a part of that commotion. Slipping quietly into the water, I watched the sun inch up on the horizon as early yellow-green rays penetrated the calm surface. Light and shadow bouncing off the craggy walls turned the coral reef into a stage set, with the actors waiting just off-stage to begin the drama. Hunters such as jacks and tuna cruised with the current, a classic stalking technique for many of the predators that patrol the reef at this hour. POW! As if a bomb had exploded, all the jacks and tuna fled. Two giant tuna burst onto the scene. They moved with incredible speed, yet their steel-silver flanks seemed almost motionless; they glided rather than swam. Working in tandem, they herded a school against a reef wall and attacked.

A hundred and fifty miles due south of Ras Mohammad, smack into the middle of the Red Sea, are The Brothers, two islands surrounded by virgin reefs. The British built a lighthouse here in the 1880s which still guides busy tanker traffic up and down the Northern Red Sea. Despite this beacon, The Brothers have claimed many ships that, over time, have been incorporated into the structure of the reef.

When I first dove The Brothers in 1982 with some intrepid Egyptian divers, I was blown away. It was an unexplored reef in rich bloom. Soft corals added pastels. Trains of schooling surgeon fish stretched out for more than a kilometer. Giant sharks and other large pelagic fish patrolled its borders.

It had then, and still does, a "magical quality," always promising the best and usually delivering on that promise.

Night diving in the Red Sea offers the thrills and surprises of exploring a city after dark.

My flashlight captures the graceful pirouettes of a Spanish dancer nudibranch in mid-water. The red eyes of poisonous scorpionfish resting up for another day of camouflage reflect the light of the hand torch. Using a flashlight lets me view the real colors and compose my photographs. In addition, it isolates my subject and forces me to concentrate on this small illuminated scene. I like to wait until the middle of night when the underwater action is at its height.

140-141 Jacks (*Carangoides cfr. fulvuguttatus*) often ride incoming tides to prey on estuarine fishes and invertebrates. Marsa Baraka, Red Sea. Depth: 15 meters.

142-143 A Napoleon wrasse (*Cheilinus undulatus*) swims together with a remora (*Echeneis naucrates*) that keep it company. Ras Nas Rani, Red Sea. Depth: 5 meters.

144-145 These shrimps *(Periclimenes sp.),* that are 2 centimeters in lenght, live symbiotically in the gill plume of the Spanish Dancer nudibranch *(Hexabranchus sanguineus).*
Ras Umm Sid, Red Sea.
Depth: 40 meters at night.

There is a better chance of capturing the animal behaviors that are the pulp of a photograph... stalking prey, a fight to the death, cannibalism, and such. In the middle hours of night, I find fish sleeping in all sorts of uncomfortable positions, offering me images that just do not occur during the day.

The Red Sea is deep in places. Diving deep is not something I recommend. Breathing compressed air under intense pressure can cause nitrogen narcosis or oxygen poisoning.

When I dive below 30 meters, I usually go with a specific purpose or image in mind.

I always work with a safety diver, but in deep diving we go over our plans and equipment with special care. My safety diver's prime task is to keep track of the time to make sure we do not exceed our margin of safety. Every second at these depths comes at great expense, so you must maximize your time. Every nerve is alert. As your eyes adjust to the world of gray-blue shadows, you search for treasures — they are everywhere. You are struck by the thought that you may be the first person to ever see this scene. The sea's strength and omnipotence possess you. Sometimes you are not quite sure whether your awe is inspired by this netherworld or by nitrogen narcosis.

146-147 Crown of Thorns *(Acanthaster planci)* starfish have a rough reputation on the reef; they feed and wipe out fields of coral. The poison spines pictured here pack a strong punch that has blown up the hands of many divers.
Eilat, Red Sea.
Depth: 8 meters.

148-149 What a face! Check out those lips and the fleshy tabs that help to give this scorpionfish superb camouflage.
Ras Umm Sid, Red Sea.
Depth: 7 meters at night.

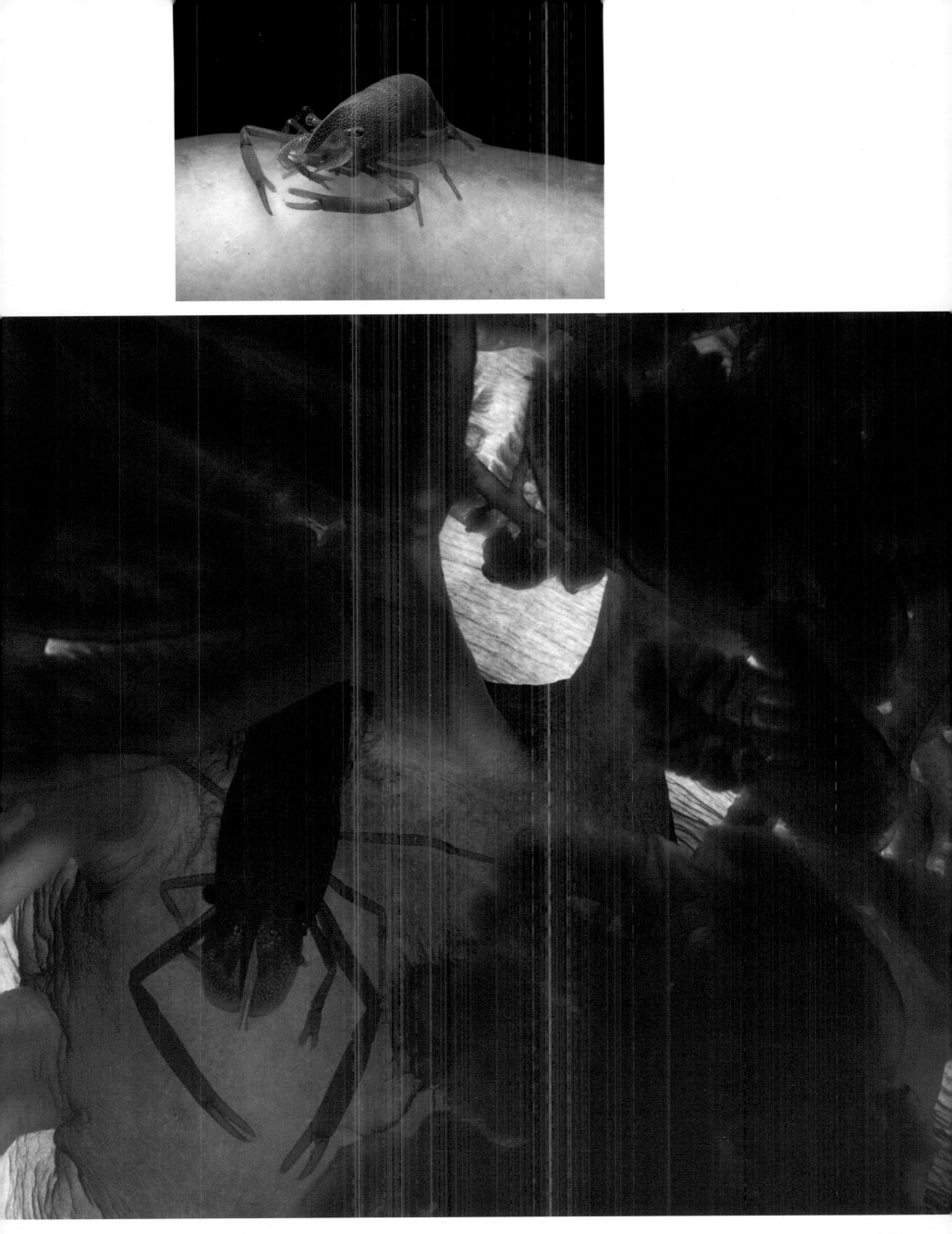

PELAGIC NIGHT

You can hardly imagine any place that seems more desolate than the open ocean at night. That is, until Bill Curtsinger, one of the best underwater photographers alive, had a brilliant idea. Far off the coast of Hawaii, in the dead of a moonless night, he switched on a car headlight suspended from a shark cage hanging off the side of his dive boat. What he saw, photographed, and later published in *National Geographic*, were like images of life from an undiscovered planet, which they were: a black, deep, pelagic world. When I called Bill, he related the whole story of his adventure, complete with a giant shark that brushed against him in 40 meters of water.

In his understated way, Bill surmised that the shark might have been a good four meters, tip to tail. When I said I couldn't wait to try his technique myself, he added, "Oh yeah... good luck!" Within a week I met with my old diving buddy from the Red Sea, Ernst Meier, a veterinarian with a sixth sense about animals and an excellent underwater photographer. Ernst immediately began work on the design and construction of a shark cage from which we would run two 1,000 watt lamps. I then enlisted David Friedman, who knows and understands the Red Sea better than anyone else alive. David suggested we begin work in April, after the deep nutrient-rich upwellings rose to the sur-

face and triggered the renewal of life in the Red Sea. The location he most favored was the Straits of Tiran with its constant current and deep, open sea. Asher Gal, a former Israeli underwater commando and professional diver, would complete the team as our safety diver.

A few weeks later, in late afternoon, our Bedouin boat captain Embarak of Sinai guided us to the deep water off the coast of Tiran Island. Far from shore, we were the only lights in an inky sea. We carefully lowered our shark cage to a depth of ten meters. We started up the generator and held our breath. POW! Two thousand watts of light knifed through the depths.

In addition to pelagic drifters that didn't migrate up and down with the light, much of what we sought were the nocturnal migrators that travel as much as a kilometer vertically toward the surface at night. We had carefully chosen a moonless night to work because we were creating our own artificial moon. At 10:00 p.m., we pulled on our wetsuits, and simultaneously, the three of us splashed backward, somersaulting into the sea. Swimming down to the cage, I fought the surge of a strong current. Once inside, I looked around me and realized our lights had drawn a blizzard of life. I put my cameras down and watched in amazement. The animals were tiny, some no larger than a pencil eraser, and completely bizarre. Just as I noticed

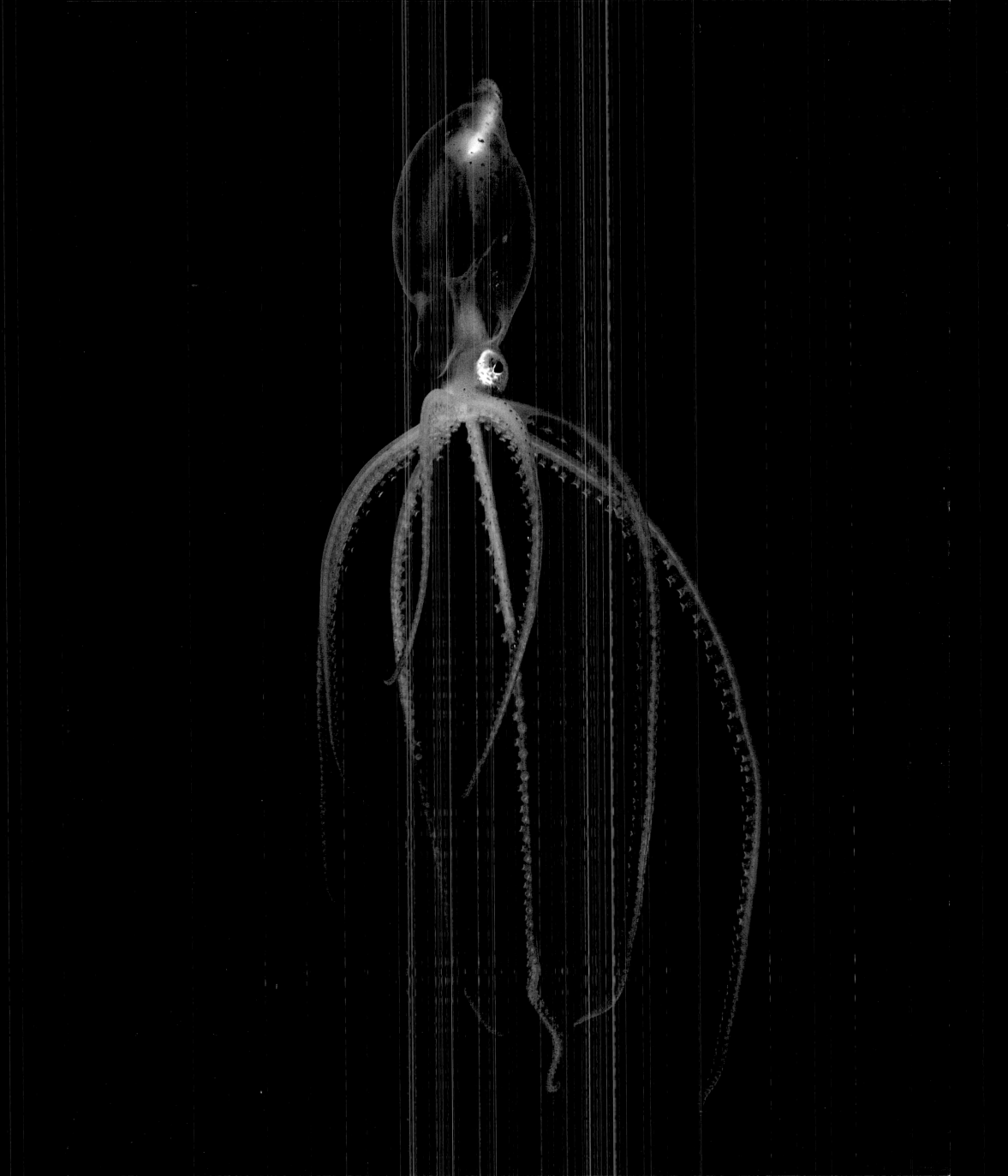

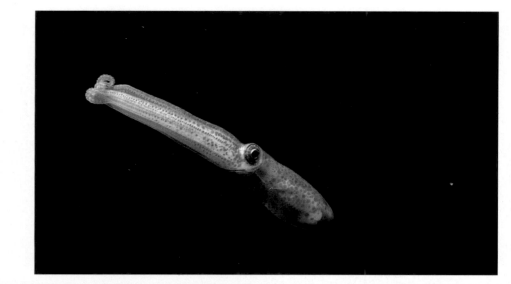

152 top At eight centimeters in length this juvenile cephalopod - possibly a pelagic type of squid manoeuvred in the open sea.
Straits of Tiran, Red Sea.
Depth: 15 meters.

153 top Migrating up from the deep night, in the open sea, this 5 centimeter-long fish picked up a hitchhiker - a tiny amphipod clings to its dorsal ridge.
Straits of Tiran, Red Sea,.
Depth: 15 meters.

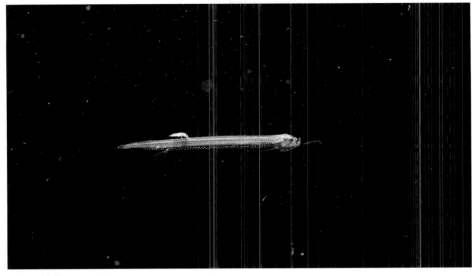

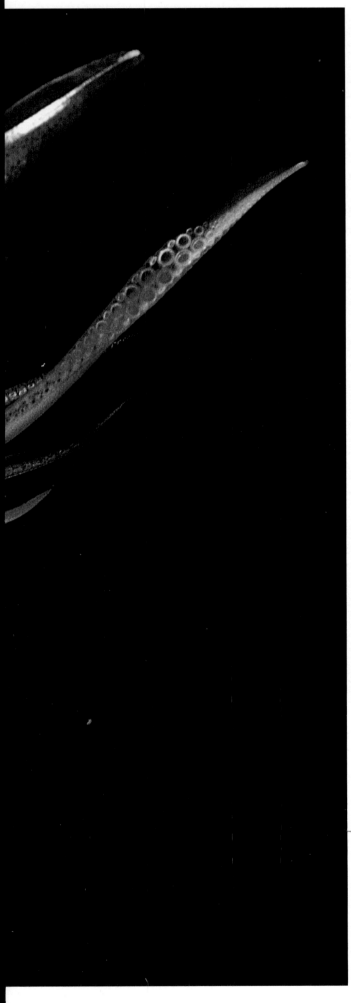

a feathery shrimp paddling toward me, a four-centimeter-long fish with enormous teeth attacked from the darkness and gobbled it up. Everywhere I looked, animals were devouring other animals in a Lilliputian feeding frenzy. No sooner would my eyes lock onto a subject than it would disappear in a gulp. I looked toward Ernst, and he acknowledged me with a smile and a thumbs-up. Asher signaled to me that we should go deeper and take a look around. Constantly glancing back at the cage for reassurance, we descended to about thirty meters. We directed our lamp beams in sweeping circles around us, aware that giant sharks were out there somewhere. I signaled Asher that I had gone into my reserve air, and we slowly ascended to the cage, collected Ernst, and climbed back on board the dive boat. We couldn't believe the incredible experience we had just had. Yelling and dancing around the deck, we probably made Captain Embarek wonder if we had finally succumbed to narcosis of the deep.

In the week that followed, our excitement never waned. Night after night, we continued to see the most bizarre animals and amazing behaviors. At 3:00 am on the last night, as we dressed for the third and final dive of the evening, Asher noticed an enormous school of fish had surrounded our cage. As I decended toward the cage, a veritable curtain of anchovy-like fish reluctantly parted to let me through. Reaching out, I grabbed a fistful of fish, but they wriggled away through my fingers. A moment later, the school broke apart and then regrouped in typical evasive action to escape a predator. But what was after them? Asher dropped down below the school, disappearing into a cloud of fish. Minutes later he came back and signaled for me to follow him. As we descended to twenty-five meters, we saw what was upsetting the school. A smaller school of about twenty jacks was circling below. Every few minutes they would break formation and rush into the school with snapping jaws. After ripping into their prey, they would back off and wait for bits and scraps of fish to float down to them. I stopped taking pictures and became a passive observer to this extraordinary event. Such an ending to the week only fueled our desire to return as quickly as possible.

We made four more trips over the next fourteen months. We experimented with different seasons and different locations. We were privileged to see some of the most extraordinary creatures of the pelagic ocean — translucent-blue pelagic octopuses, larval fishes of all types, deep sea fish with symbiotic amphipods attached to their bodies, gelatinous jellies that reflected every shade of the rainbow with their beating cilia. The pelagic waters had shared its secrets.

152-153 Dropping down to 20 meters I spotted this pelagic squid-10 centimeters in length a somewhat large animal in the drifting broth that makes up open water plankton. At 40 meters I froze it momentarily with my flash long enough to produce this image drifting off.
South Sinai, Red Sea.
Depth 40 meters.

154 What is this? These
are the words that came to
my mouth when I found
this animal drifting in open
sea at midnight. Making a
circle the size of an egg
this type of posture seems
defensive. It is possibly a
new species which could
be in a family of
anguilliphorms-a specialist
in fish larvae is needed for
positive identification.
Straits of Tiran,
Red Sea.
Depth: 20 meters.

155 An open water,
pelagic shrimp drifts as
part of the planktonic
community on a
moonless night.
Straits of Tiran,
Red Sea.
Depth: 10 meters.

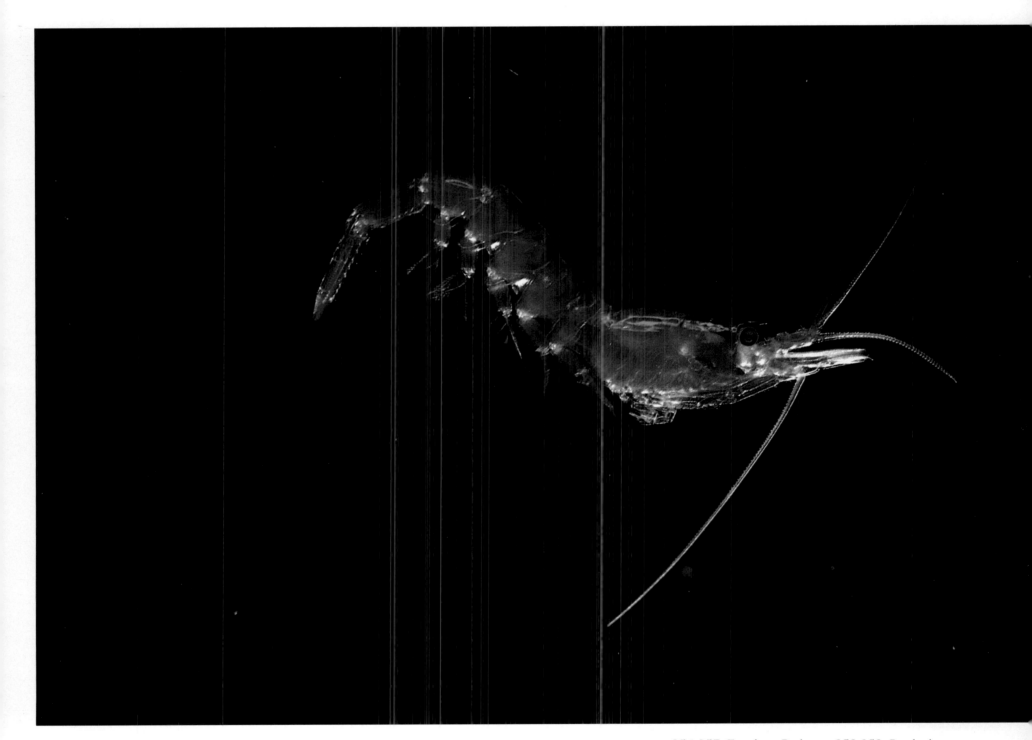

156-157 For these Red
Sea glassy sweepers
(Pempheris sp.), the
anonymity and confusion
of large schools provide
protection from predators.
Brothers Islands, Red Sea.
Depth: 10 meters.

158-159 Perched on a
living ledge conveniently
provided by its host
anemone, an anemone
shrimp *(Periclimenes sp.)*
eyes the camera. Warily?
Defensively? Nervously?
It's hard to tell with
shrimp.
Ras Nas Rani, Red Sea.
Depth: 5 meters.

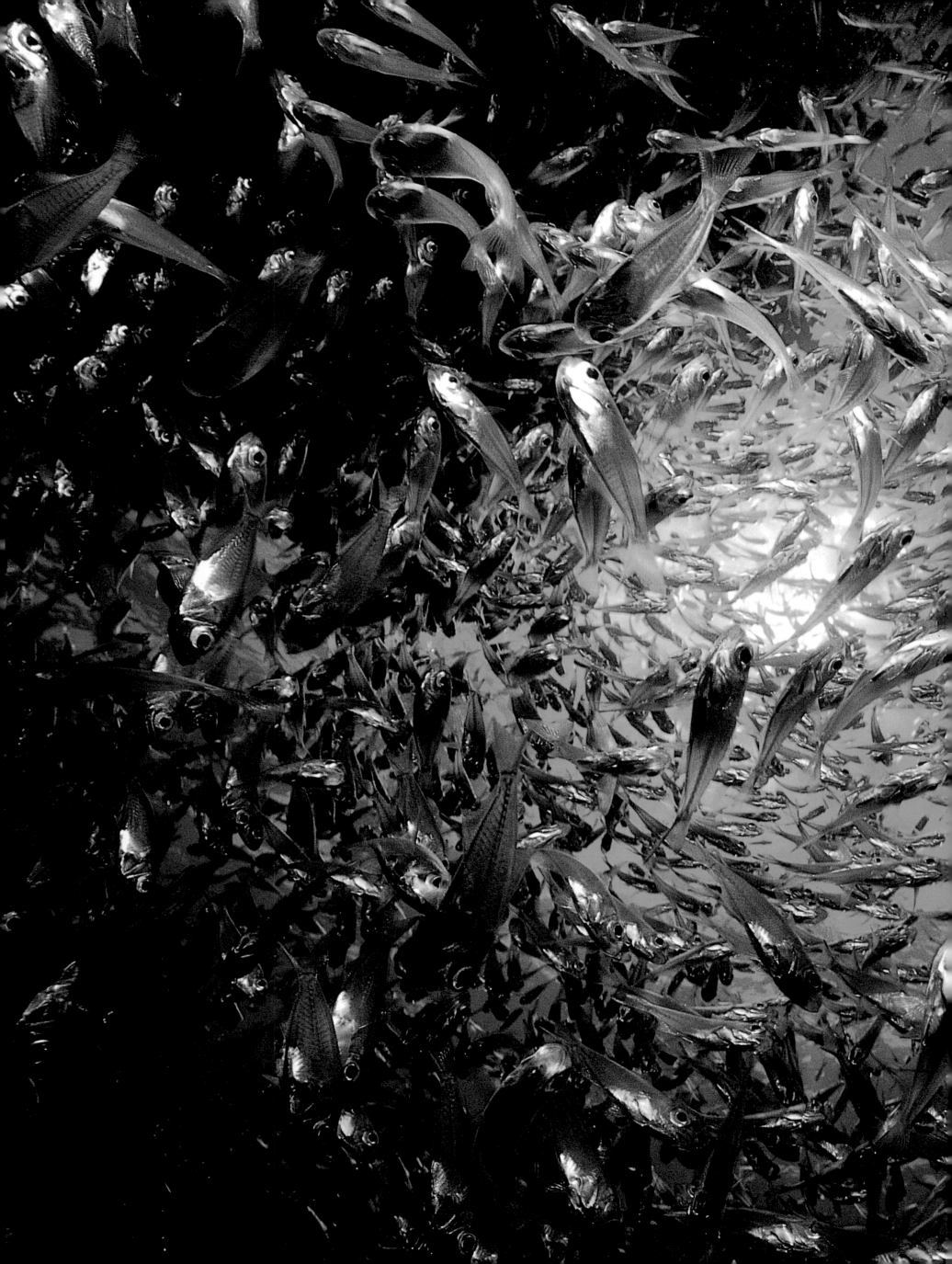

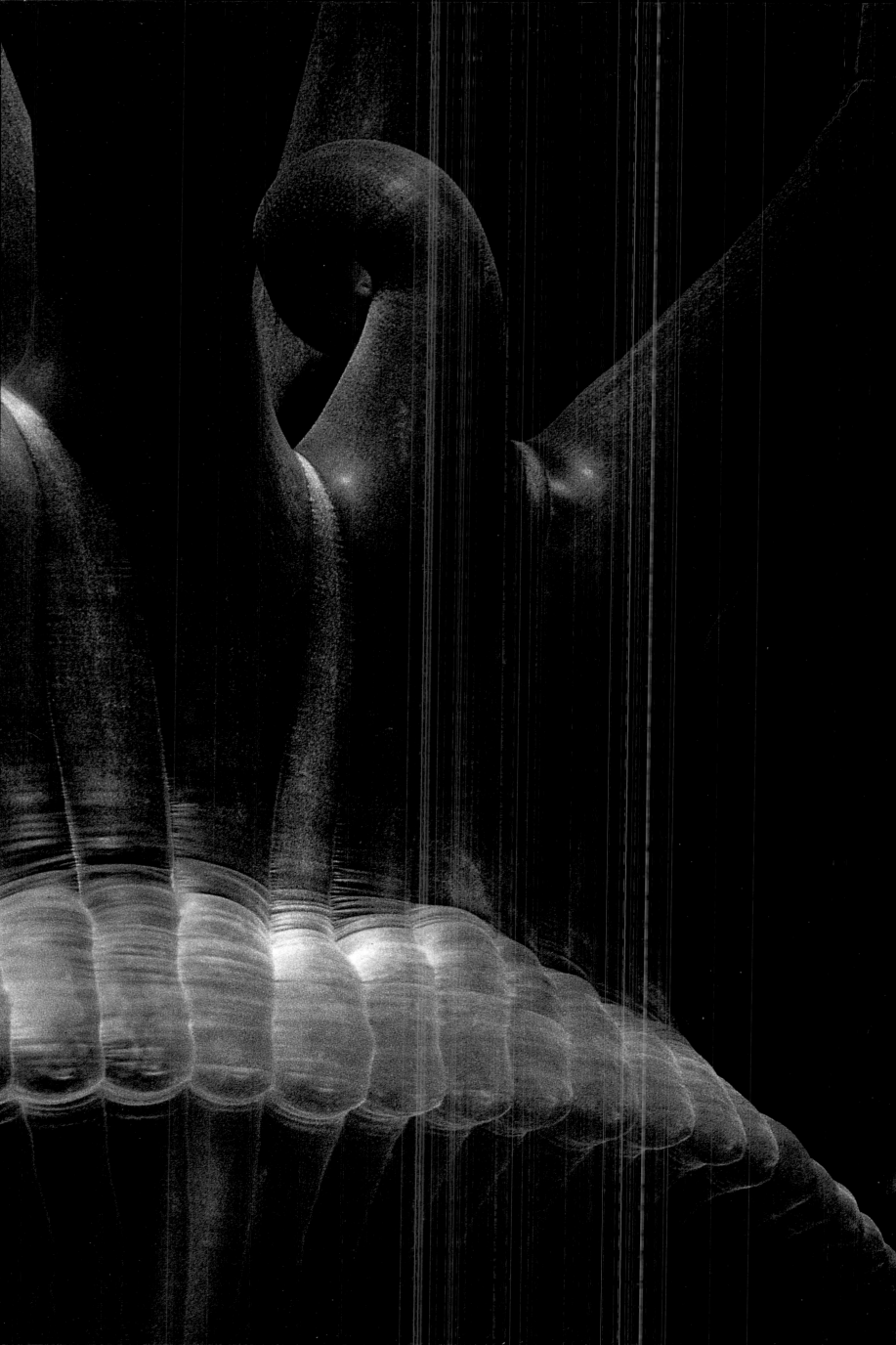

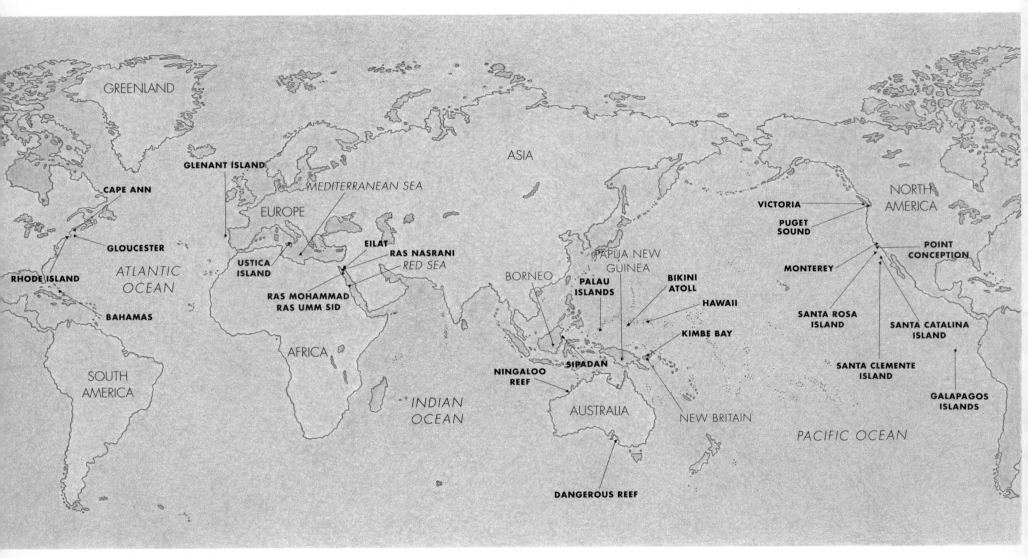